THE EVOLUTION OF CYBER WAR

The Evolution of Cyber War

International Norms for Emerging-Technology Weapons

BRIAN M. MAZANEC

POTOMAC BOOKS
An imprint of the University of Nebraska Press

 Potomac Books is an imprint of the University of Nebraska Press.
Manufactured in the United States of America.

Library of Congress Cataloging-in-Publication Data
Mazanec, Brian M., author.
The evolution of cyber war: international norms for emerging-technology weapons / Brian M. Mazanec.
pages cm
Includes bibliographical references and index.
ISBN 978-1-61234-763-9 (hardback: alk. paper)
ISBN 978-1-61234-774-5 (epub)
ISBN 978-1-61234-775-2 (mobi)
ISBN 978-1-61234-776-9 (pdf)
1. Information warfare (International law)
2. Cyberspace operations (Military science)
3. Cyberterrorism—Law and legislation.
4. Technological innovations—Law and legislation.
I. Title.
KZ6718.M39 2015
341.6'3—dc23
2015008309

Set in Minion by Westchester Publishing Services

To Charlotte, Reagan, Peter, and Benjamin:

Thomas Paine once wrote,

"If there must be trouble, let it be in my day, that my child may have peace."

May you live in an age of peace.

CONTENTS

ILLUSTRATIONS

GRAPH

TABLES

ACKNOWLEDGMENTS

An effort such as this is not possible without the significant support from many individuals. I would like to thank my editor at the University of Nebraska Press, Marguerite Boyles, for her support of this project and for guiding it to success.

I would also like to thank the many individuals who directly and indirectly contributed to this work, including Penney Harwell Caramia, Joe Kirschbaum, Greg Koblentz, Angelos Stavrou, Brad Thayer, Trevor Thrall, and Matt Ullengren.

Additionally, I would like to thank my parents, Dan and Polly, for their unending love and support and whose words and examples instilled in me an appreciation of the value of education.

Most important, I would like to thank my incredible wife, Abby, for her unwavering support of this undertaking. Her love, encouragement, and patience were indispensable, and her scholarly work ethic served as an inspiration throughout this project. I would also like to thank my four awesome children—Charlotte, Reagan, Peter, and Benjamin—who sacrificed time and provided essential motivation to see this work to completion.

Any errors herein are mine. Additionally, the views expressed herein are mine and do not represent the views of any organization or entity I am affiliated with.

ABBREVIATIONS

BBSU	British Bombing Survey Unit
BW	biological weapons
BWC	Biological Weapons Convention
CBM	confidence-building measure
CBW	chemical and biological weapons
CIA	Central Intelligence Agency
CMA	Chemical Manufacturers Association
CNA	computer network attack
CNE	computer network exploitation
CSIS	Center for Strategic and International Studies
CTBT	Comprehensive Test Ban Treaty
CTBTO	Comprehensive Test Ban Treaty Organization
CTITF	Counter-Terrorism Implementation Task Force
CW	chemical weapons
CWC	Chemical Weapons Convention
CYBERCOM	Cyber Command
DDOS	distributed denial-of-service
DOD	Department of Defense
EWI	East-West Institute
FAS	Federation of American Scientists
GGE	group of government experts
IAEA	International Atomic Energy Agency
ICS	Industrial Control Systems
IDC	International Data Center
IMS	International Monitoring System

IP	Internet protocol
ISP	Internet service provider
ISR	intelligence, surveillance, and reconnaissance
IT	information technology
ITU	International Telecommunication Union
LOAC	laws of armed conflict
LTBT	Limited Test Ban Treaty
NAM	nonaligned movement
NAS	National Academy of Sciences
NATO	North Atlantic Treaty Organization
NATO CCD COE	Cooperative Cyber Defence Centre of Excellence
NDAA	National Defense Authorization Act
NGO	nongovernmental organization
NIC	National Intelligence Council
NNWS	nonnuclear weapon states
NPT	Nuclear Nonproliferation Treaty
NRC	National Research Council
NSA	National Security Agency
OPCW	Organization for the Prohibition of Chemical Weapons
PGM	precision-guided munitions
PLA	People's Liberation Army (China)
PLC	programmable logic controller
PNET	Peaceful Nuclear Explosions Treaty
RAF	Royal Air Force
REVCON	review conference
RMA	revolution in military affairs
SALT	Strategic Arms Limitation Talks
SCADA	Supervisory Control and Data Acquisition
SROE	standing rules of engagement
TTBT	Threshold Test Ban Treaty
UNIDIR	United Nations Institute for Disarmament Research
USSBS	U.S. Strategic Bombing Survey
WMD	weapons of mass destruction
WPC	World Peace Council

Introduction

In the conclusion of his book *The History of a Crime*, Victor Hugo wrote, "One resists the invasion of armies; one does not resist the invasion of ideas."[1] This statement, made in reference to the ideals of the French Revolution, alludes to the fact that ideas, when spread, can grow, shape, and dominate. One category of ideas is known as norms, which are shared expectations of appropriate behavior.[2] Norms exist at various levels and apply to different actors. In the international arena, these nonbinding shared expectations can, to some degree, constrain and regulate the behavior of international actors and, in that sense, have a structural impact on the international system as a whole. For example, early in the age of nuclear weapons, Lt. Gen. James Gavin expressed the contemporary wisdom when he wrote, "Nuclear weapons will become conventional for several reasons, among them cost, effectiveness against enemy weapons, and ease of handling."[3] However, as the nuclear era advanced, a constraining norm developed that made states reluctant to possess or use nuclear weapons. Views similar to those held by Gavin and others at the dawn of the nuclear era regarding military utility and inevitable employment also existed with strategic bombing at the advent of the ability to conduct aerial bombing of civilians during wartime in the early 1900s.[4] These permissive views were once held regarding chemical and biological weapons.

International security and U.S. national security may be enhanced by the emergence of some kind of regulative norm for cyber warfare, similar to norms that developed in the past for these other emerging-technology weapons. In March 2013 Director of National Intelligence James Clapper testified that a major cyber attack against the United States could occur as soon as in just the next two years, resulting in "long-term, wide-scale disruption of services, such as a regional power outage." He further stated that the growing international use of these emerging-technology weapons to achieve strategic objectives was outpacing the development of a shared understanding or norms of behavior and thus increasing the prospects for miscalculations and escalation.[5] Today, early into the age of cyber warfare, many hold a view regarding the inevitability of significant use of force in cyberspace similar to that held early in the nuclear era. In expectation that norms will emerge, in May 2011 the Obama administration issued the *International Strategy for Cyberspace*.[6] One pillar of this strategy recognizes the "borderless" international dimension of cyberspace and identifies the need to achieve stability and address cyber threats through the development of international norms. In February 2013 Michael Daniel, the White House cyber security coordinator, told computer security practitioners that diplomacy, including fostering international norms and shared expectations, is essential to prevent cyber warfare against U.S. economic interests, and in March 2013 National Security Advisor Tom Donilon called for China to agree to "acceptable norms of behavior in cyberspace."[7] This book seeks to explain how constraining norms for cyber warfare—along with other emerging-technology weapons—are developing and will develop in the future by refining norm evolution theory for emerging-technology weapons. To do so, it develops a norm evolution theory for emerging-technology weapons based on case studies on the evolution of norms for other emerging-technology weapons—specifically chemical and biological weapons, strategic bombing, and nuclear weapons.

Key Concepts

Norms are standards of right and wrong that form a prescription or proscription for behavior.[8] Essentially norms are nonbinding, shared expectations that can be helpful in constraining and regulating the behavior of international actors and, in that sense, have a structural impact on the international system. International norms cover a wide range of issues, from the practice of dueling to human rights. Specific to warfare, multiple regulative norms have emerged regarding specific categories of weapons and modes of warfare, such as weapons of mass destruction (WMD), strategic bombing, antipersonnel land mines, leadership assassination, and dueling.[9] As discussed in more detail in chapter 1, norms for weapons and warfare can affect a variety of functions and activities, such as weapon possession and use. While not always successful (with the demise of the constraining strategic bombing norm in World War II being perhaps one of the better examples), some of these norms for warfare have helped restrain the widespread development, proliferation, and use of various weapons. Norms are obviously one of many variables impacting state behavior and the international system, with material state interests and power dynamics also playing a major, perhaps dominant role. As discussed in more detail in chapter 1, this book does not seek to measure the relative impact of norms compared with these other factors but merely explains how they emerge and develop.

Although in January 2014 Pope Francis called the Internet a "gift from God," there is not a universal or infallible consensus on key terms and definitions regarding cyberspace.[10] Within the United States, the cyber domain is defined by the U.S. Department of Defense (DOD) as the global realm within the "information environment" consisting of the interdependent network of information technology infrastructures, including the Internet, telecommunications networks, computer systems, and embedded processors and controllers.[11] Some argue that the full electromagnetic spectrum should be included in any definition of cyberspace, which would make electronic warfare such as radar jamming a form of cyber attack. However such a definition is extremely broad, and most experts have a more limited view. Cyberspace operations are the employment of cyber capabilities where the primary purpose is to achieve objectives in or through cyberspace.[12] The range of definitions of cyber conflict can be as broad as Alan Campen's 1995 definition of cyber conflict as conducting, or preparing to conduct, military operations according to "information-related principles" wherein the emphasis of the operation is more on the disruption of critical information than historically conventional targets.[13] One subset of cyber conflict is cyber warfare—another term lacking a universally agreed upon definition but generally used in reference to computer network exploitation (CNE) and computer network attack (CNA). CNE is the more frequent of the two and entails the use of computer networks to gather intelligence on an adversary.[14] On the more violent end of the spectrum, CNA is the use of computer networks to disrupt, deny, degrade, or destroy either the information resident in enemy computers and computer networks or the computers and networks themselves. This understanding of cyber warfare, which is focused on CNA between state actors (directly or through plausibly deniable nonstate clients) is the focus of this book. Recent examples of cyber warfare were seen in Estonia in 2007, Georgia in 2008, and Iran in 2010, all of which contained elements of both CNE and CNA.[15] As with other forms of warfare, cyber warfare targeting can be countervalue (focused on civilian targets) or counterforce (focused on military personnel, forces, and facilities). The general lack of consensus on what constitutes cyber warfare and even cyberspace itself highlights some of the challenges facing the emergence of constraining norms for cyber warfare. Additional background information on cyberspace and cyber warfare is found in the appendix.

Research Question, Approach, and Central Argument

This book explains how constraining norms for cyber warfare—along with other emerging-technology weapons—are developing and will develop in the future by refining norm evolution theory for emerging-technology weapons. To do so, it examines norm evolution theory and then tests it in three case studies of emerging-technology weapons. Beyond being emerging-technology weapons, these case studies—chemical and biological weapons, strategic bombing, and nuclear weapons—are in some respects similar to cyber warfare. The development of constraining norms for cyber warfare may be able to advance more quickly by recognizing and adopting approaches that have been learned through efforts to encourage the evolution of norms for other emerging-technology weapons. To approach this research question in this way, norm evolution theory—discussed in detail in chapter 1—is used as an analytic framework to help develop primary and secondary hypotheses for norm evolution for emerging-technology weapons, including the identification of the specific actors, actor motives, and important mechanisms and factors in the case studies that supported norm emergence and development throughout all stages of the norm life cycle. Yogi Berra once said that "in theory there is no difference between theory and practice; in practice there is."[16] The case studies bridge these two worlds and ground norm evolution theory in practice by helping to both test the theory and refine it specifically for emerging-technology weapons, which will then be used to assess how similarly constraining norms for cyber warfare may develop. These three main case studies are covered in chapters 2 (chemical and biological weapons), 3 (strategic bombing), and 4 (nuclear weapons) through a review of the existing weapon-type-specific literature. Chapter 5 offers a refined version of norm evolution theory, tailored for emerging-technology weapons. Chapter 6 applies this new norm evolution theory of emerging-technology weapons to cyber warfare, first by reviewing the current status of cyber warfare norms (discussed in more detail in the appendix) and then by applying this refined theory developed from the case studies to anticipate and predict how norms for emerging-technology cyber weapons may develop from their current state should trends continue. Finally, some brief conclusions and recommendations are offered regarding constraining norms for cyber warfare and future research related to the evolution of norms. Ultimately this book argues that for emerging-technology weapons, direct or indirect alignment of national self-interest with a constraining norm is the primary factor that leads to norm emergence, and the extent to which it is aligned with key or powerful states perception of self-interest will determine how rapidly and effectively the norm emerges. It also identifies secondary variables and hypotheses that support norm emergence and growth during each stage of the norm life cycle. Specific to cyber warfare, while multiple candidate norms are beginning to emerge through state practice and deliberate norm cultivation efforts by an increasing number of organizational platforms, norm evolution theory for emerging-technology weapons indicates that, if current trends continue, constraining norms for cyber warfare may not successfully emerge and may never reach a norm cascade. Should the constraining norms emerge successfully, the odds of reaching a tipping point are better, although internalization is less likely. Potential contingency scenarios, such as the occurrence of a major cyber attack, could alter the prospects of cyber warfare norm evolution. Of the current candidate cyber norms identified in chapter 6, the most likely to succeed are those that are more limited in scope, such as those focused on applying the existing laws of armed conflict (LOAC) to cyber warfare or prohibiting first use of cyber weapons. In light of these findings, this book concludes by recommending that the United States (1) pursue only limited consensus on cyber norms and not completely constrain development of offensive cyber capabilities, (2) judiciously employ said capabilities and offer cyber extended deterrent to allies to manage proliferation, (3) seek limited norm cultivation, and (4) manage the prospect for miscalculation in cyberspace by increasing transparency and confidence-building measures. Additionally, this book argues that further research is warranted to continue to develop norm evolution theory for emerging-technology weapons, as such a theory has broad implications for norm evolution across a range of new weapon types as we head deeper into the twenty-first century. Specifically researchers should seek to further validate and refine norm evolution theory for emerging-technology weapons through additional case studies of emerging-technology weapons and also seek better mechanisms to measure the relative impact of norms.

Case Studies

Chemical and biological weapons—at one time emerging-technology weapons—and cyber weapons are forms of nonconventional weapons that share many of the same characteristics with significant international security implications.[17] They include challenges of attribution following their use, attractiveness to weaker powers and nonstate actors as asymmetric weapons, use as a force multiplier for conventional military operations, questionable deterrence value, target and weapon unpredictability, potential for major collateral damage or unintended consequences due to "borderless" domains, the multiuse nature of the associated technologies, and the frequent use of covert programs to develop such weapons.[18] Due to these characteristics, these weapons are attractive to nonstate actors and those seeking lack of clarity regarding the responsible party. Because of these common attributes, lessons regarding norm development can be learned from chemical and biological weapon experiences that are applicable to nascent efforts to address the use of cyber weapons. Some chemical and biological warfare norms are codified in contractual obligations and binding agreements such as treaties, as was the case with biological weapons when the Convention on the Prohibition of the Development, Production, and Stockpiling of Bacteriological (Biological) and Toxin Weapons and on Their Destruction, commonly referred to as the Biological Weapons Convention, or BWC, entered into force in 1975. The BWC codified the existing norm against development, production, and stockpiling of biological weapons and declared their use to be "repugnant to the conscience of mankind." The BWC now has 155 state parties. Earlier norms against use of these weapons led to the 1925 Geneva Protocol's prohibition on their first use in a conflict. (States retained their right to retaliate with such weapons.) Other binding agreements codifying these norms exist, including the Chemical Weapons Convention (CWC), which prohibits outright all chemical weapons. Examining the factors leading to these successes is helpful in developing a framework to predict how constraining norms for cyber weapons may evolve.

Strategic bombing—particularly with the advent of airpower as an emerging-technology weapon and the early use of airplanes to drop bombs on cities—forced states to grapple with a brand-new technology and approach to warfare, as is now the case with cyber warfare. As with chemical and biological weapons, strategic bombing shares some characteristics with cyber warfare. Strategic bombing made civilian populations highly vulnerable, was difficult to defend against, and used technology that also had peaceful applications (air travel and transport)—all of which can be said about cyber warfare today. At the end of the nineteenth century technology had advanced to the point where substantial aerial bombing of civilian and military targets from balloons was conceivable. Such strategic bombing, particularly of civilian targets, appeared to conflict with the existing norm of noncombatant immunity. At the Hague Peace Conference of 1899 the participants agreed to prohibit the "discharge of explosives or projectiles from balloons" for a period of five years. The codification of this emerging-technology weapon norm struggled through various debates before, during, and after World War I. However, by the 1930s there was consensus that bombing civilians was unacceptable, even drawing an admission from Adolf Hitler in 1935 that a "prohibition on indiscriminate bombing of densely populated regions" was warranted.[19] This norm collapsed during World War II. It eventually reemerged from the ashes of the conflict and developed into an enduring norm. The effort to constrain strategic bombing through normative influences was mixed and at times completely unsuccessful, which makes it particularly well suited as an exemplar of the limits of norms and how other factors may impede or reverse norm development.

Nuclear weapons, like airpower earlier and perhaps cyber weapons today, presented states with the challenge of a completely new war-fighting technology. Nuclear weapons and cyber weapons, like the other emerging-technology case studies, share many of the same characteristics with significant international security implications. These include the potential for major collateral damage and unintended consequences (due to fallout, in the case of nuclear weapons) and covert development programs. While early nuclear norms were permissive and did not constrain the United States from deploying nuclear bombs on Hiroshima and Nagasaki, that soon changed. As noted by Thomas Schelling, the rapid emergence of norms against the use of nuclear weapons was so effective in constraining action that President Eisenhower's secretary of state, John Foster Dulles, when contemplating the use of nuclear weapons in 1953 (less than a decade after the first use of nuclear weapons), said that "somehow or other we must manage to remove the taboo from the use of [nuclear] weapons."[20] This constraining nuclear norm was eventually internalized and codified in agreements such as the Nuclear Nonproliferation Treaty, which enacted limits on nuclear proliferation and a commitment to eventual disarmament. Examining the successful emergence, cascade, and internalization of the constraining nuclear norm may help point a path to success with prospective cyber norms.

1 General Norm Evolution Theory

> One resists the invasion of armies; one does not resist the invasion of ideas.
>
> Victor Hugo, *The History of a Crime*

Cyber warfare poses significant challenges to U.S. national security. As one response to these challenges, the United States issued an "International Strategy for Cyberspace" in May 2011.[1] This strategy identifies the need to achieve stability and address cyber threats through the development of international norms. So, what are international norms? This chapter attempts to answer this question and provide a theoretical underpinning for how norms emerge, spread, and ultimately impact state behavior. First, norms are introduced, with a discussion of how they are considered in the context of various international relations theories. Norm evolution theory and the norm life cycle are also discussed. The chapter then applies norm evolution theory to identify the specific hypotheses concerning each phase of the norm life cycle. Norm evolution theory—grounded in extensive literature from fields such as international relations, sociology, economics, evolutionary biology, organizational theory, and other disciplines—will provide the foundation for later chapters' discussions of norms regarding emerging-technology weapons, such as chemical and biological weapons, strategic bombing, nuclear weapons, and, of course, cyber warfare.

Norms are standards of right and wrong that form a prescription or proscription for behavior.[2] Norms are considered one component of regimes, which are "principles, norms, rules, and decision-making procedures around which actor expectations converge in a given issue-area" and thus constrain some actor behaviors.[3] In other words, norms—which are nonbinding—are shared expectations or "standards of appropriateness" and can exist at various levels and apply to different actors.[4] At their most basic, norms exist at the community level, where they apply to the behavior of individuals. For example, community norms governing sexual behavior, such as the expectation that sexual intercourse is appropriate only between married couples, have a constraining impact on individuals in the community where the norms reside.[5] A community norm may not be held by all individuals, but its existence influences everyone's behavior and raises the risks and benefits associated with engaging in behavior discouraged (in this case, premarital sexual intercourse) or encouraged by the norm. A community may be a small subset of a regional population, such as a particular religious or ethnic group, or may consist of the entire regional population. In addition to the community level, norms can influence organizations, including governments, at the national level as well as state behavior overall at the international level. At the national level, norms can have a structural impact on internal domestic behavior. As a perhaps extreme example, in Saudi Arabia the strong Islamic culture and resulting norms support the existence and acceptance of a religious "morality" police force, officially referred to as the Committee for the Promotion of Virtue and the Prevention of Vice, which, among other things, enforces religious norms requiring a separation of the sexes in public.[6] Beyond the national level, these nonbinding shared expectations can, to some degree, constrain and regulate behavior of international actors and, in that sense, have a structural impact on the international system as a whole. Often international norms emerge from national norms. Ronald Jepperson, Alexander Wendt, and Peter Katzenstein look at the development of international norms through what they call "sociological institutionalism."[7] This essentially means that norms, as part of a global political culture, impact our national security environment by actively shaping views regarding issues such as national sovereignty, international law, and diplomacy. This dynamic of constantly changing norms influences the character of states and acceptable state behavior. For example, norms regarding nuclear weapons have had a pronounced impact on state behavior. Early in the age of nuclear weapons, Lt. Gen. James Gavin expressed the contemporary wisdom when he said, "Nuclear weapons will become conventional for several reasons, among them cost, effectiveness against enemy weapons, and ease of handling."[8] However, as the nuclear era advanced, a constraining norm developed that made states more reluctant to possess or use nuclear weapons, thus helping prevent their "conventionalization." Nobel Laureate Thomas Schelling noted that this rapid emergence of norms against the use of nuclear weapons was so effective in constraining action that President Eisenhower's secretary of state, John Foster Dulles spoke of the need to remove the "taboo" from the use of nuclear weapons so that they could be used.[9] This norm was so strong that President Truman did not use nuclear weapons against Chinese troops during the Korean War, President Nixon did not use them in Vietnam, the

Soviet Union did not use them in Afghanistan, and Israel did not use them in the 1973 war with Egypt—all circumstances where the opponent lacked such weapons and their use made some degree of utilitarian sense. This normative tradition against the use of nuclear weapons was developed and influenced by the "logic of consequences," a realist and rationalist argument regarding self-interest and the negative outcomes nuclear weapon users would experience, and the "logic of appropriateness," which reflects the fact that the nuclear norm itself actually influenced states' identity and was reinforced and deemed appropriate through iteration over time.[10] While both pure realists and constructivist advocates of the role of nonmaterial normative factors in international relations have difficulty quantifying the comparative explanatory power of norms, the example of norms for nuclear weapons demonstrates that they have some impact. International norms similar to this nuclear norm have emerged regarding other forms of unconventional weapons, such as biological and chemical weapons and strategic bombing.

In addition to categorizing norms as applying to various levels (community, national, and international), norms can generally fall into two broad categories: constitutive and regulative norms. Constitutive norms "create new actors, interests, or categories of action" and do not create any particular duty or permission. In contrast to constitutive norms, regulative norms (sometimes also referred to as deontic norms) "order and constrain behavior" and actually have some kind of overt prescriptive quality.[11] An example of a constitutive norm is the institution of marriage wherein men and women form a particular type of bond. Another example is the emergence of a national symbol such as a song becoming a national anthem, changing the very meaning of singing that song. An example of a regulative norm is the view in the 1920s and 1930s that submarine attacks against merchant ships were heinous and immoral.[12] Another example is the norm against the employment of nuclear weapons. Constitutive norms can also enable regulative norms. For example, the institution of marriage must exist before an expectation of being married under certain conditions or before engaging in certain behaviors can emerge. This book is focused on regulative norms rather than constitutive norms since those are the types of norms that most directly influence conflict between states and apply to particular weapons, including novel and unconventional weapons such as cyber weapons. Regulative norms can be constraining or permissive and can also apply to a nearly limitless range of behaviors. For emerging-technology weapons, a constraining norm would be one that indicated that doing something with a weapon (possessing it, using it, etc.) was not acceptable, while a permissive norm would obviously indicate the inverse. In addition to being constraining or permissive, norms for weapons and warfare generally affect four general activities or functions: weapon development or possession, testing, use, and transfer or proliferation facilitation. These "flavors" of norms are depicted in table 1.

In this book, which is largely focused on the evolution of constraining rather than permissive norms, multiple norms focused on some or all of these particular activities or functions for a given weapon type (for example, nuclear weapons) are bundled together and considered collectively as an overall set of norms for that weapon type. For example, when referring to the constraining "nuclear norm" or "norm for nuclear weapons" in chapter 5, unless otherwise specified this reference is to the comprehensive set of specific norms affecting these four activities or functions for the weapon.

Table 1. Activities or functions affected by norms for weapons and warfare

ACTIVITY OR FUNCTION	NORM DESCRIPTION
Weapon development or possession	Programs to develop a particular weapon are acceptable or unacceptable, and possession of said weapons is acceptable or unacceptable.
Weapon testing	Testing in general, or certain types of testing (atmospheric, etc.) of a particular weapon is acceptable or unacceptable.
Weapon use	Use or certain types of use (first use, use against nonmilitary targets, etc.) of a particular weapon is acceptable or unacceptable.
Weapon transfer or proliferation facilitation	Efforts to transfer or facilitate proliferation of a particular weapon are acceptable or unacceptable.

Norms and International Relations Theory

Scholars view norms differently depending on the international relations theory they espouse. Realists and liberal internationalists place less emphasis on normative factors over material state-power calculations, while Grotian internationalists and constructivists lend more credence to the role of norms.

Realists and Norms

The concept and role of norms in broader international relations theory have been the subject of debate for some time. Adherents to the Hobbesian or realist conception of international relations are focused on matériel power considerations and not necessarily ethical or normative concerns, and they may be inclined to discount or dismiss the role of norms in governing international relations. Realists generally view the international system as anarchic and in a constant state of zero-sum competition and see states as the only meaningful actor. Historically realists do not pay much attention to norms. After all, if one believes that, as Thucydides said, "the strong do what they will, the weak accept what they must," there appears to be little room for norms to constrain or influence the strong.[13] Realists often view norms as peripheral reflections of the dominant states seeking to maximize their interest.[14] The hegemonic stability theory also explains this view when it argues that states are forced to adhere to norms or adopt norm-based regimes because of coercion from a hegemonic state.[15] However, some strains of realism, such as neoclassical realism, have begun to allow room for normative factors. In the narrowest sense, neoclassical realists allow for ethical and normative arguments regarding how states may define their interests (based on domestic culture, obligations, notions of right and wrong, and so on).[16] Some of these neoclassical realists have begun to acknowledge that some (but not all) statesmen follow ethical norms and that in the anarchic and competitive international system powerful states may be self-constrained from doing what they can (for example, acquiring WMD, antipersonnel land mines, or possibly conducting cyber warfare) in order to comply with a normative construct that provides a net benefit by also constraining adversaries and weaker powers.[17]

Liberal Internationalists and Norms

Like realists, adherents to liberal internationalist theories of international relations have historically viewed norms only as minor variables in the international system. Liberal internationalists also view the state as the primary actor; however, they view the international system as mostly good or capable of being good, and thus mutually beneficial international cooperation is possible. Because of this generally positive view regarding international cooperation, liberal internationalists are more inclined to believe norms matter insofar as they arise from voluntary cooperation among states. However, they view the role of norms as limited to only this voluntary (and possibly infrequent) cooperation and not as a major factor for individual state behavior or actions.[18] The theory of complex interdependence, first introduced by Robert Keohane and Joseph Nye's book *Power and Interdependence: World Politics and Transition* in 1977, is an offshoot of the liberal internationalist school of international theory but also incorporates some realist tendencies. Essentially, complex interdependence views the international system as anarchic but increasingly interdependent. Complex interdependence refers to "a situation among a number of countries in which multiple channels of contact connect societies (that is, states do not monopolize these contacts); there is no hierarchy of issues; and military force is not used by governments toward one another."[19] Such a perspective allows more room for normative factors as they can be viewed as part of the connections binding societies. Complex interdependence theorists believe that, in addition to nation-states, there are a variety of subnational, international, transnational, and supranational actors who can share a variety of economic, social, and political connections, including normative (ideational) connections. According to Keohane and Nye, eventually complex interdependence leads to globalization, which will minimize the use of force. This globalism will be "a state of the world involving networks of interdependence at multi-continental distances, linked through flows and influences of capital and goods, information and ideas, people and force, as well as environmentally and biologically relevant substances (such as acid rain or pathogens)."[20] Such a view allows norms a greater role than that traditionally granted by realists, but it is still a relatively small variable in the theory.

Grotians, Constructivists, and Norms

It is the Grotian (also known as internationalist) and constructivist theories of international relations that grant the most emphasis to norms as a variable that can influence international behavior. Constructivists believe the way individuals look at the world affects what they see, and thus many of the structural constants perceived by realists and liberal internationalists (such as the anarchic system and primacy of state actors) are largely governed by malleable cultural assumptions. Thus constructivists believe the international system depends on "the eye of the beholder" and places ideas and identity politics in a central role.[21] However, constructivists' rejection of a positivist view of the world and their resulting emphasis on unique aspects of each situation significantly limit their ability to offer an empirically verifiable structural theory for international relations. In contrast, the internationalist conception of the international system does not completely reject the state-focused structural theory of international relations; however, it allows much more room for the consideration of nonmaterial factors such as norms when compared to realism or liberal internationalism. Internationalists view states as primary actors but posit that they operate in a "society of states" that provides constraining norms and principles that influence state behavior. Hedley Bull explained this concept: "The particular international activity which, as the Grotian view, best typifies international activity as a whole is neither war between states, nor horizontal conflict cutting across the boundaries of states, but trade—or, more generally, economic and social intercourse between one country and another."[22] In the context of this "social intercourse," internationalists believe that norms have a significant and independent impact on the international system and shape state behavior.[23] They do not reject the idea that states are driven by their calculations of interest and power, only that they are also influenced by the international society (that is, the normative environment) in which they reside. In fact internationalists view norms and state calculations or interest and power as mutually conditioning.[24]

While norms are viewed differently depending on the international relations theory employed, for the purposes of this book it does not matter if norms are viewed as major variables that can sometimes trump or constrain state-power calculations or if they play a lesser role. Even many realists now acknowledge the role of nonmaterial factors in international affairs, which demonstrates that examining norms for how various weapons develop and grow is not only a constructivist or internationalist pursuit.[25] Rather than attempting to settle this broader theoretical debate, this book seeks to offer a theory of norm evolution for emerging-technology weapons, especially cyber warfare.

Norm Evolution Theory

There is a wide-ranging and interdisciplinary literature that discusses the emergence and development of international regulative norms. Norms have been utilized as a lens for understanding international activity with increasing frequency, due in part to behavioral and microeconomic research lending support to the tangible role of norms.[26] This norm evolution theory is largely focused on the emergence and development of norms and not so much their relative impact compared to other variables such as state interest and material factors. Ann Florini introduced an evolutionary analogy based on natural selection to explain how international norms change over time.[27] Natural selection, introduced by Darwin in *On the Origin of Species*, is the gradual, nonrandom process by which biological traits thrive or perish in a population. Norms too can thrive or perish. For example, the initially strong norm against strategic bombing eroded and ultimately perished due, in part, to "inadvertent escalation" resulting from strategic bombing's compatibility with the war-fighting culture of each nation's military services during World War II; the increasing desperation of the state actors and their calculation that the moral opprobrium wrought by violating the norm had become secondary to the existential benefit of using such weapons; and pivotal technological change, such as the invention of the Norden bombsight and improved inertial navigation that made strategic bombing more militarily effective.[28] Natural selection entails variation in traits, differentiation (or selection) in reproduction, and replication through hereditary genetics.[29] This evolutionary approach to norms contributes significantly to the theory of norm emergence and development by helping explain why particular norms change over time.[30] Norms, like genes, are instructional units that influence the behavior of their host organisms. For regulative norms, their phenotype is the particular behavior they address.[31] Genes and norms are both transmitted through inheritance; in the case of norms it is from one state to another (horizontal reproduction) or internally within a state (vertical reproduction).[32] Vertical norm reproduction refers to a continuation of a norm through leaders in a single state, and norms reproduced in this way rarely change. In contrast, horizontal norm reproduction is diffusion across multiple states in a single generation. It is this type of norm reproduction that is most relevant when considering norms governing weapon technology and warfare as such norms need to be spread across multiple states in order to influence state-to-state conflict.

Overall, norm evolution theory identifies three major stages in a norm's potential life cycle: (1) norm emergence, (2) norm cascade, and (3) norm internalization.[33] Collectively these life cycle stages cover the full spectrum of norm evolution, from the nascent emergence of a novel norm to its nearly total adoption and codification across the globe. However, a norm may never move through all three stages and can reach its terminal development at any stage and possibly even regress and dissipate.

Stage 1: Norm Emergence

Norms, like genes, are contested, meaning there is competition with other norms, and as a result norm prevalence in a population changes over time. Different norms have different levels of "reproductive advantage," a different likelihood of being transmitted.[34] Norm emergence is the stage wherein a new norm initially comes into existence, often tenuously. Norm evolution theory—borrowing from biological evolutionary theory—identifies several specific conditions that are essential for a new "mutant" or emergent norm to replace an existing one. For example, a norm currently emerging is the norm intended to constrain the illicit spread of small arms and light weapons such as submachine guns, rifles, automatic weapons, and light machine guns.[35] This small arms norm, intended to create expectations regarding weapon possession and transfers, is not yet established and must actively be supported in order to emerge. The first condition needed for norm success at this stage is whether a norm is prominent enough in the existing "norm pool" to gain a "foothold."[36] This concept is similar to the idea of gene prominence that often occurs when a population is geographically isolated and able to separate from the "noise" of the environment that can overwhelm small mutations. For norm emergence, that prominence usually occurs when a particular actor is energetically supporting and promoting the norm, referred to as a "norm entrepreneur." Norm entrepreneurs arise as agents and advocates of a particular norm and use organizational platforms as a mechanism to begin changing existing norms and promoting the new norm. Martha Finnemore and Kathryn Sikkink identify Henry Dunant, a Swiss banker who helped found what ultimately became the Red Cross, as an example of a strong norm entrepreneur. Following his experience at the battle of Solferino in 1859, he successfully advocated for the norm that medical personnel and the wounded be treated as neutrals and noncombatants during conflict. These individual norm entrepreneurs are often motivated by factors such as altruism, empathy, or ideological commitment and primarily use persuasion as a mechanism to promote their norm, although they may engage in disruptive behavior such as civil disobedience in order to advocate for their norm in a contested normative environment.[37] Norm entrepreneurs either call attention to the new norm they are promoting or even create it through reinterpretation or "framing" of the issue. Thomas Schelling highlights this process in regard to nuclear warfare norms by discussing the work of international dialogue (facilitated through organizations like the Aspen Institute and the Institute for Strategic Studies in London) in spreading and promoting the norm against nuclear use.[38] An interesting and yet unclear dynamic is the emergence of nonstate actors that could be considered "anti-norm-entrepreneurs," that is, the emergence of nonstate actors such as terrorists who completely ignore prominent international norms. William Potter has explored how nuclear norms may be changing with the rise of these terrorist actors who seek nuclear weapons and are not directly influenced or constrained by existing nuclear norms. However, he argues, perhaps counterintuitively, that terrorist use of nuclear weapons may not erode—and could perhaps enhance—the norm against nuclear weapon use, in a way ironically making those nonstate actors who contravene the norm inadvertent norm entrepreneurs.[39]

Another key element for norm emergence is the previously mentioned organizational platform norm entrepreneurs use to promote their norms.[40] Organizational platforms can be a nongovernmental organization (NGO) or a standing international organization such as the World Bank. Different organizational platforms provide different types of resources and advocacy networks that can be used to secure the support of state actors. As an example, scholars have identified the importance of organizational platforms by highlighting the role of NGOs such as Human Rights Watch, Handicap International, Physicians for Human Rights, Medico International, Mines Advisory Group, and the Vietnam Veterans of America Foundation in the effort to establish the emerging norm against antipersonnel land mines. [41] Through these norm entrepreneurs and organizational platforms, norm emergence largely builds as some states (considered norm leaders) internally respond and adopt norms.

In addition to norm prominence and the existence of entrepreneurs and organizational platforms, norm evolution theory indicates that various additional factors can fuel or stifle a norm's growth and development during the emergence stage of the life cycle. These include specific mechanisms used by certain actors, such as persuasion and framing, overarching environmental factors associated with specific states or the entire international community, and some factors associated with the emerging norm itself. Norm leaders often respond to and adopt norms for domestic political reasons. For example, domestic political concerns were the driving factor behind codifying norms regarding human rights.[42] Further, Nina Tannenwald argues that the norm regarding nuclear weapons arose through an unintentional and fortuitous grassroots genesis (that is, from the bottom up), meaning it emerged from views held by the general public who then influenced their domestic political leaders.[43] States insecure about their international status and seeking international "legitimation" are more likely to adopt emerging norms. Eventually enough states adopt the new norm and a "tipping point" is reached.[44] Evidence of this tipping point can be seen as the norm begins to become institutionalized in specific international rules and organizations, although such institutionalization may not fully occur until after the tipping point is reached and a norm cascade has occurred.

Connections to certain ideas, such as a prohibition of bodily harm to innocent bystanders or making an "adjacency" claim that connects the norm to existing norms can help a new norm take root.[45] The term *coherence* is used to describe how successfully a new norm interacts with existing and widespread norms with which it is not directly competing.[46] Others have referred to it as *congruence* with the preexisting local, regional, and international normative order.[47] Coherence is similar to the idea of grafting onto existing norms, and it helps lend legitimacy to a nascent norm. For example, the NGO-led campaign to prohibit antipersonnel land mines deliberately sought to graft their nascent norm against antipersonnel land mines to the widespread prohibition on indiscriminate killing of civilians due to the adverse impact of millions of old and forgotten antipersonnel land mines worldwide.[48] Making these adjacency claims may require proactive construction of connections and ideational bridges by norm entrepreneurs. The literature on strategic culture also highlights how social context and strategic culture can impact leadership or population receptivity to budding norms (and, by extension, social construction by norm entrepreneurs), with some cultures less receptive to normative influences than other cultures.[49] Florini points out that this cultural transmission facilitated by norm entrepreneurs occurs primarily through mechanisms such as "memes" or metanarratives.[50] A meme is an "idea, behavior, style, or usage that spreads from person to person within a culture" and can include things like a normative catch-phrase or a particular fashion.[51] For example, with the effort to establish a norm against antipersonnel land mines, additional tools such as the use of persuasive memes and shame helped the emerging norm take root.[52] Antipersonnel land mine opponents advocating for the Mine Ban Treaty and norms against antipersonnel land mines and cluster munitions employed specific advertisements in developed, mine-free nations with themes such as "If there were land mines here, would you stand for them anywhere?" and a video of a young girls' soccer match in the United States during which a land mine explodes and injures players.[53] Another framing mechanism used by norm entrepreneurs to advance their emerging norm is shifting the burden of proof. An example of this—again based on the experience of the emerging antipersonnel land mine norm—is the use of a proselytizing advocacy network set up by the NGOs to foster a "transnational Socratic method," which forces nations retaining antipersonnel land mines and resisting the norm to publicly justify their actions, essentially shifting the burden of proof. This shaming and shifting of the burden onto the holdouts led to the rapid emergence and success of the antipersonnel land mine norm in the 1990s.[54] Tools such as these allow norm entrepreneurs to help attach the new norm to existing and widely accepted norms—in this case with the norm against indiscriminate killing of civilians. These tools and the concepts of coherence and grafting generally convey additional legitimacy on new norms as they become viewed as a "reasonable behavioral response to the environmental conditions facing the members of the community."[55] In a sense, this means emerging norms thrive when they are viewed as logical extensions or necessary changes to existing norms.

Various items associated with the environmental conditions facing the norm pool can further bolster the success of an emerging norm. This can include the specific states involved as norm leaders and their environmental factors, such as the distribution of power, prevailing levels of technology, and availability of natural and human resources.[56] For the effort to establish the norm for antipersonnel land mines—in addition to grafting and framing the issue effectively—environmental factors such as the overall geopolitical conditions (that is, the absence of major conflict or wars), the effects of globalization, and the attitude of other actors in the international system were major elements of its success as an emerging norm.[57] Unfortunately, some, such as the scholar Diana O'Dwyer, believe the convergence of all these factors is difficult to replicate. Also the large-scale turnover of decision makers (through revolutions or other internal events), the obvious failure of norms from a previous norm generation, and the emergence of new issue areas in which prevailing norms are not yet well established can further establish a fertile environment for a new norm to emerge and grow.[58] However, specific to norm evolution theory for emerging-technology weapons, the presence of some of these environmental factors may actually be counterproductive for norm emergence. This is due to uncertainty regarding the new weapons themselves and associated technological constraints. For example, the efforts to establish a norm against strategic bombing of civilian targets during the interwar period was unsuccessful in part due to the challenge of avoiding indiscriminate collateral damage due to a lack of precision.[59] However, this challenge was not permanent, and improvements in bombing technology in the 1980s and 1990s with precision-guided munitions (PGMs) have reinvigorated the strategic bombing norm as it has made such an approach feasible while retaining the ability to bomb military targets.

The hypotheses that norm evolution theory offers regarding the actors, actor motives, and important mechanisms and factors supporting norm emergence and growth during the emergence stage of the norm life cycle are summarized in table 2.

Stage 2: Norm Cascade

Once a tipping point has been reached, the second stage in the norm life cycle, a norm cascade, begins. A norm cascade occurs when a norm's international adoption significantly accelerates and states adopt the norm in response to international pressure regardless of the presence of a domestic coalition advocating for the norm. Norm scholars caution, however, that there is no way to specify a priori when a tipping point will be reached or precisely how many states will need to take up a norm for it to crest over this critical threshold. That said, empirical research indicates that generally a norm tipping point is rarely reached without at least one third of the total number of states in the international system adopting the norm.[60] However, the specific states adopting the norm sometimes matter more than the total number of states involved. Some states have more "moral stature" or are otherwise considered critical to the norm due to their role in whatever the norm affects (for example, states that use or produce antipersonnel land mine for any norm regarding such weapons). Finnemore and Sikkink point out that once the tipping point is reached and the norm cascade begins, international norm adoption begins at a much more rapid pace. No longer is norm adoption largely a function of domestic movements and pressure; instead it spreads through "international socialization." At this stage norms are more likely to spread and be internalized if they are held by states viewed

Table 2. Hypotheses for norm emergence

NORM EMERGENCE	
Actors	Norm emergence is more likely to occur when key actors are involved, specifically: • Norm entrepreneurs with organizational platforms • Norm leaders (states) Norm emergence is more likely to occur if more actors are involved.
Actor motives	States and key actors are likely to support and participate in norm emergence when motivated by various factors, specifically: • Altruism • Empathy • Ideational commitment
Important mechanisms and factors	Norm emergence is more likely to occur when important mechanisms and factors are involved, specifically: *Actor mechanisms* • Persuasion—framing, memes, shame *Environmental factors* • Domestic political concerns • States seeking international legitimation • State distribution of power, prevailing levels of technology, availability of natural and human resources, etc. favoring the norm • Large-scale turnover of decision makers • Recognized failure of norms from a previous norm generation • Emergence of a new issue area where prevailing norms not well established *Norm factors* • Prominence • Coherence—connections to key ideas, adjacency, and grafting

Source: Table developed in part from Martha Finnemore and Kathryn Sikkink, "International Norm Dynamics and Political Change," *International Organization* 52, no. 4 (1998): 898.

as desirable models—an extension of the concept of norm prominence introduced earlier. During this norm cascade, states respond to this international socialization for a variety of reasons, but most often legitimating, conformity, and esteem.[61] States that are norm leaders will often use bilateral or multilateral diplomatic praise or censure (sometimes to include material pressure such as sanctions) to encourage norm adoption. Additionally, NGOs and nonstate advocacy networks can apply social pressure. Increased international interdependence and rapid communication resources have spawned increased opportunities for norm socialization. Sometimes this socialization is done to advance a particular state's or group of states' interests. For example, norms constraining chemical weapons initially developed to complement the interests of the major powers. However, over time the memories of these state interests fade and the norm remains as an independent force.[62] During this socialization process, as a norm cascade grows, persuasion and framing continue to play a major role. For example, with WMD-related norms, internationally categorizing certain weapons as "unconventional" and "weapons of mass destruction" contributed to the development of norms constraining their use.[63] Additionally, during this norm cascade stage, optimal timing due to a shock or major international event can help push a norm to this tipping point or accelerate its internalization.[64] For example, the norms associated with the losing side in a war or with an economic collapse leading to a depression may be rapidly discredited and alternative norms sought.[65] The current globalized international environment can also be considered an aspect of the optimal timing, and some have argued that it has accelerated the prospects for some norms moving through all three stages of the norm life cycle. For example, the women's suffrage norm took 120 years to move through the first two stages in an environment lacking globalization, while the norms associated with violence against women progressed to the same degree in only twenty years due to increased global interdependence.[66]

The hypotheses that norm evolution theory offers regarding the actors, actor motives, and important mechanisms and factors supporting norm emergence and growth during the cascade stage of the norm life cycle are summarized in table 3.

Stage 3: Norm Internalization

The third and final stage in the norm life cycle (which may never be reached) is norm internalization. This phase is the effort to deeply codify the norm and achieve universal or near universal adherence.[67] At this stage, the norm is so deeply and broadly internalized that it achieves taken-for-granted status. Norm scholars point out that this can actually make the norm hard to discern because all debate and controversy surrounding it may be resolved.[68] However, achieving internalization does not necessarily mean the norm is internalized 100 percent in every state, and the degree to which it is internalized can vary across states and regions. Also internalization does not necessarily mean the norm is fully codified in binding and near universal international agreements or treaties, although that may occur. As with the earlier stages, various internal and external factors contribute to a norm's movement in this stage. For example, Andrew Moravcsik looks at the adoption and emergence of norms relative to human rights and identifies domestic political concerns as the catalyst for deeply internalizing norms regarding human rights.[69] Specifically norms regarding human rights were adopted by nascent democracies out of concerns that nondemocratic forces may later gain power and that they want to "lock in" norms in order to constrain

Table 3. Hypotheses for norm cascade

NORM CASCADE	
Actors	A norm cascade is more likely to occur when key actors are involved, specifically: • States • International organizations • Transnational networks A norm cascade is more likely to occur if more actors are involved.
Actor motives	States and key actors are likely to support and participate in a norm cascade when motivated by various factors, specifically: • Legitimacy • Reputation • Esteem
Important mechanisms and factors	A norm cascade is more likely to occur when important mechanisms and factors are involved, specifically: *Actor mechanisms* • Persuasion—diplomacy • Institutionalization • Socialization • Demonstration *Environmental factors* • Optimal timing *Norm factors* • Clear and specific • Universal claims • Coherence

Source: Table developed in part from Martha Finnemore and Kathryn Sikkink, "International Norm Dynamics and Political Change," *International Organization* 52, no. 4 (1998): 898.

these potential future domestic actors. However, generally speaking, domestic factors play a lesser role at this stage than they do in the earlier stages.[70] External factors can include the norm's role in shaping identities. For example, the nuclear norm, while still not fully internalized, has become associated with the idea of what constitutes a "civilized nation."[71]

Other identified factors that can contribute to successful norm internalization are incorporation into professions and iterated behavior and habit.[72] Finnemore and Sikkink identify professions as a powerful tool for socializing individuals to various norms; once a norm is built into the professional training its internalization is greatly facilitated.[73] Additionally, frequently repeated behaviors in governmental bureaucracies and relevant professions (connected in some way to the norm) leading to habits help further institutionalize the norm over time. This phenomenon is referred to as *institutional isomorphism* in sociology's "new intuitionalism" literature, indicating that institutions begin to mimic each other over time and with routines and repetition (deepening the norm's institutionalization).[74] Again, as with the previous stages, the intrinsic characteristics of the norm are also factors that can help explain why or why not a norm spreads and is internalized. Norms that are both clear and specific (rather than ambiguous and complex) or make universal clams rather than localized ones are generally more likely to have a deep and lasting transnational impact.[75] For example, the late nineteenth- and twentieth-century efforts to develop norms for acceptable behavior of airpower were stymied due to the complexity of the prospective norms and associated competing interests and viewpoints. This complexity was grounded in the fact that precision targeting was not yet possible and civilian aircraft could be easily converted to military purposes, and thus the proposed norms against possessing such weapons were not practically effective.[76] While the Hague Peace Conference of 1899 resulted in a five-year prohibition on aerial bombing, that was as far as this norm advanced at this time due to intrinsic characteristics associated with the norm.[77] Soon after World War II broke out, any lingering sense of a norm restricting strategic bombing deteriorated rapidly.[78] Additionally, certain norms have a particular coherence (as in the earlier stages) with the overall environment that facilitates their internalization. For example, some argue that the norm against chemical weapons is so well internalized because it is grounded in long-standing and near universal primitive human experience with poisonous snakes and associated ancient mysticism.[79] This is reflective of the same ideas popularized by the anthropologist Mary Douglass, who examines the concepts of pollution and taboo and their normative effects on culture.[80]

Now that general norm evolution theory has been introduced, the hypotheses it offers regarding the actors, actor motives, and important mechanisms and factors supporting norm emergence and growth during the internalization stage of the norm life cycle are summarized in table 4.

While norm evolution theory is useful for predicting and explaining the development and evolution of international norms, it was in large part developed using case studies unrelated to warfare, such as social issues involving women's rights, protecting the environment, or encouraging free trade.[81] As such, it is not specifically developed nor tailored for norms for weapons and warfare, particularly those involving emerging technology. As a result of the broad and discursive approach used to develop norm evolution theory, it consists of an expansive list of hypotheses (see table 4), and with such a broad list of variables it is not sufficiently discriminating or targeted so as to

Table 4. Hypotheses for norm internalization

NORM INTERNALIZATION	
Actors	Norm internalization is more likely to occur when key actors are involved, specifically: • Law and lawyers • Professions • Bureaucracy Norm internalization is more likely to occur if more actors are involved.
Actor motives	States and key actors are likely to support and participate in a norm cascade when motivated by various factors, specifically: • Conformity • Domestic turmoil
Important mechanisms and factors	Norm internalization is more likely to occur when important mechanisms and factors are involved, specifically: *Actor mechanisms* • Habit • Institutionalization *Environmental factors* • Existence of professional networks *Norm factors* • Clear and specific • Universal claims • Coherence

Source: Table developed in part from Martha Finnemore and Kathryn Sikkink, "International Norm Dynamics and Political Change," *International Organization* 52, no. 4 (1998): 898.

best facilitate predictions of norm evolution for emerging-technology weapons, such as those for cyber weapons. Norm evolution theory's expansive list of hypotheses and variables puts it in danger of tautology. Jeffrey Legro points out some of these deficiencies as well as the norm literature's general focus on showing that norms "matter" to the detriment of developing more detailed and specific theories.[82] In light of these weaknesses, while norm evolution theory will be applied to the selected case studies in the following three chapters in order to generally validate the broad theory, it will also be refined and tailored to make it more useful for emerging-technology weapons.

2 Norm Evolution for Chemical and Biological Weapons

> We have established . . . international norms that say when you use these kinds of weapons, you have the potential of killing massive numbers of people in the most inhumane way possible, and the proliferation risks are so significant that we don't want that genie out of the bottle.
>
> President Barack H. Obama

As discussed in chapter 1, biological and chemical weapons (BW and CW, respectively—collectively referred to as chemical and biological weapons [CBW]) and cyber are forms of nonconventional weapons that share many of the same special characteristics and have significant international security implications. In the first half of the twentieth century, CBW could be considered emerging-technology weapons, as cyber weapons are today. Accordingly, examining norm evolution for CBW is a valuable case study to help extend norm evolution theory and develop expectations for how norms for cyber warfare will develop. This chapter examines the norm life cycle of CBW and the various hypotheses from norm evolution theory.

Along with nuclear weapons (examined in chapter 4), CBW are often categorized as WMD. Biological weapons are pathogenic microorganisms, such as bacteria and viruses, or toxic substances of biological origin used to kill, sicken, or disorient an enemy.[1] Chemical weapons are toxic chemicals also used to kill, sicken, or disorient an enemy.[2] Both types require a delivery system, such as a bomb, crop duster, or living vectors (for example, mosquitoes, birds, bats) in order to achieve their effects. Based on limited knowledge and understanding of biology and microorganisms, these two weapon types were initially combined as "poisons" without categorical distinction. For example, the 1925 Geneva Protocol dealing with BW and CW referred to them collectively as "asphyxiating, poisonous or other gases, and . . . all analogous liquids, materials or devices and bacteriological methods of warfare."[3] Successful development of norms against the possession and use of CBW have helped constrain their proliferation and use, and today they are nearly universally considered "repugnant to the conscience of mankind" (as stated in the 1972 BWC, which now has 155 state parties).[4] In the modern era a significant number of major and widespread attacks involving BW and CW have not occurred. Major CW attacks occurred only during World War I, the regional war between Egypt and Yemen in the 1960s, the Iran-Iraq War and lingering conflict in Iraq in the 1980s, and in 2013 in the Syrian Civil War (although these last two incidents elicited a strong international response).[5] The only confirmed battlefield use of BW against humans was by Imperial Japan against China in the 1930s and 1940s.[6] The BWC and CWC ultimately banned this entire class of weapons, with the CWC including strong verification measures. So how did this CBW norm emerge, reach a norm cascade, and ultimately achieve internalization and near universal adherence? Overall the hypotheses of norm evolution theory are consistent with the experience of the constraining CBW norm's emergence, cascade, and internalization. In general, two categories of regulative norms arose: those regarding the use of BW and CW and those regarding development, possession, and proliferation of these weapons. Among the factors norm evolution theory identifies as important, the CBW norm's coherence with the long-standing poison taboo, the support of powerful and self-interested actors, optimal timing, and the involvement of key professions played critical roles. The CBW norm experience offers some additional insights to complement the general hypothesis of norm evolution theory, including the impediments and challenges that secrecy and multiuse technology offer to norm evolution and the fact that norms regarding use, especially first use, may be easier to internalize than those regarding development, proliferation, and disarmament.

Biological and Chemical Weapon Norm Emergence

Norm emergence is the stage wherein a new norm initially comes into existence, often tenuously. Norm evolution theory identifies several specific conditions that make norm emergence more likely, specifically the number and type of actors involved, important mechanisms these actors can employ to promote the norm (such as persuasion and framing), and various environmental and norm factors that can further support norm emergence (distribution of power, prevailing levels of technology, norm coherence, and so on). The emergence of constraining norms for chemical and biological weapons was consistent with these various hypotheses from norm evolution theory, with coherence with the long-standing and prominent poison taboo playing a pivotal role. This helped the modern CBW norm rapidly emerge and spread and quickly generated the involvement of key actors and norm entrepreneurs.

Coherence and Grafting of CBW *Norms with the Poison Taboo*

There is extensive human history of taboo associated with poisons, which lays a contextual foundation for thinking about chemical and biological weapons.[7] Accordingly, unlike some norms, the norm emergence stage for CBW is not quite a linear progression, due in part to an innate, perhaps biological internalization of the foundational norm on which the CBW norm was based. Coherence and grafting are important mechanisms and factors identified by norm evolution theory. Consistent with this theory, the modern CBW norm developed largely as a result of its coherence with and grafting onto these historic and poison taboos. More than anything else, the coherence of modern CBW norms with this long-standing and prominent poison norm accounts for the new norm's successful emergence.

The CBW norm may have a basis in biological evolution and natural selection, as is believed to exist for behaviors such as cannibalism and incest. An aversion to poisonous and toxic substances may have a genetic basis, as indicated by research in the field of sociobiology.[8] This idea appears logical as a natural aversion to harmful substances would improve the human species' chances of survival and reproduction; scientists speculate that some phobias (evoked by rats, spiders, snakes, and so on) are rooted in this basis.[9] Other scholars, such as Leonard Cole, disagree with this notion and instead find the basis of the norm rooted in a conscious decision of those in power to exclude "an indefensible weapon from the contestation of power."[10] Cole dismisses the genetic foundation argument as "so strained and implausible as to not merit serious consideration."[11] While Cole has a valid point that the genetic argument is only a theory and more research is needed, historical evidence from human civilizations across the globe indicate that most considered poisons—explicitly or implicitly—inappropriate weapons of war.[12] This is not due to a lack of familiarity with the concept of poisons as weapons. Human history offers plenty of examples of people firing poisoned arrows or, more rarely, catapulting plague-infected cadavers at an enemy.[13] For example, during the 1346 Siege of Caffa, a port city on the Crimean peninsula in the Black Sea, the attacking Tartars experienced an outbreak of plague and then hurled the infected bodies of their fallen soldiers over the walls of the city. The residents of Caffa, fleeing to Italy, carried the plague with them and thus furthered the second major epidemic of Black Death in Europe.[14] However, whether technically BW or CW and regardless of the delivery system, poison weapons were nearly universally considered "insidious, subtle, and sneaky." Hindu commentaries from as far back as 600 BC express this disdain for those who employ poison weapons, and the Hindu Laws of Manu expressly prohibit the use in battle of weapons that are "concealed, barbed, or smeared with poison or whose points blaze with fire." Both the Greeks and Romans were equally scornful of poison weapons; in 82 BC the Roman ruler Sulla promulgated the *Lex Cornelia*, which provided particularly severe penalties for those who used poison weapons. Conversely, poison weapons were frequently used for hunting animals, so the lack of use was not due to a lack of capability. In addition to potential biological origins, Cole attributes this ancient norm against poisons to the mystery associated with medicine (which can enhance life) and poison (which can end it), as well as snakes.[15] Fear of snakes is common not only in humans but also many other primates. The earliest human writings describe snakes and their poisonous venom as a symbol of evil. For example, the Book of Genesis uses a snake as the very personification of evil in the Garden of Eden.[16] Ultimately no matter the source—biology, culture, or both—the aversion to and related constraining norm against poison has proven to be nearly universal and provided the foundation for norm entrepreneurs to graft coherent CBW norms.

As a result of humanity's long-running taboo against the use of poison, obtaining a norm foothold did not require the same level of energetic norm entrepreneurs, organizational platforms, or norm leaders as the earliest form of the norm was already engrained in most human societies. This also meant that actor motives such as altruism and empathy and mechanisms such as persuasion and framing were less relevant during the norm emergence stage. The key factor for norm emergence was the overwhelming prominence of the poison taboo, and this norm factor of prominence and coherence with the human condition (regardless of whether it was culturally or biologically driven) provided all that was needed for the CBW norm to take root. The real story begins with the norm cascade and norm internalization stages of the norm life cycle for CBW.

Biological and Chemical Weapon Norm Cascade

A norm cascade occurs when a norm's international adoption significantly accelerates and states adopt the norm in response to international pressure regardless of the presence of a domestic coalition advocating for the norm. Norm evolution theory identifies several specific conditions that increase the likelihood of achieving a norm cascade: the number and type of actors involved (including international organizations and transnational networks), important mechanisms these actors can employ to promote the norm (institutionalization, socialization, and so on), and various environmental and norm factors that can further support norm emergence (optimal timing, clear and specific norms that make universal claims, and so on). Given the CBW norm's coherence with and success in grafting onto the widespread poison norm, actors (often powerful and self-motivated) were able to quickly use institutionalization and socialization across various transnational networks to spread the norm. In addition to coherence with the historic poison taboo, the norm's early success was due to its coherence with the humanitarian notion of noncombatant immunity. However, this foundation proved inadequate in World War I, and the norm backslid as its humanitarian underpinnings collapsed. Following the war, institutionalization and socialization (sometimes unintentional) resumed, and legitimacy, reputation, and esteem motivations maintained the norm during World War II.

Powerful Self-Interested Actors and Transnational Networks

In the late 1800s the deeply rooted CBW norm began to reach a tipping point, and a norm cascade began to occur. It was not so much that the norm's international adoption accelerated (as it was largely diffused throughout human societies), but the norm was clarified and began to deepen in advance of internalization. By the late nineteenth century the stigma associated with poison weapons had been discussed in legal and governmental avenues in Europe for a few hundred years, with an absolute prohibition being first discussed between the fifteenth and eighteenth centuries.[17] During this time Hugo Grotius published *The Law of War and Peace*, in which he said that the poison norm "arose from a consideration of the common advantage, in order that the dangers of war, which had begun to be frequent, might not be too widely extended. And it is easy to believe that this agreement originated with kings, whose lives are better defended by arms than those of other men, but are less safe from poison, unless they are protected by some respect for law and by fear of disgrace."[18] Grotius identifies the self-interested leaders (who would not be immune from the effects of CBW) as a major factor in reaching the norm cascade. This is not a traditional actor motive identified by norm evolution theory (such as legitimacy, reputation, and esteem), but it was an important factor early in the norm cascade. However, the motives identified by norm evolution theory became more relevant later in the norm cascade and during the norm internalization stage.

Humanitarianism and Other Factors Facilitated Institutionalization

Later in the modern era, the CBW norm reached the apex of its norm cascade due to a variety of factors, many identified by norm evolution theory. Norm evolution theory indicates that a norm cascade is more likely to occur when transnational networks and early international organizations exist, persuasive diplomacy is used to promote the norm, coherence with existing norms and ideas is achieved, and legitimacy incentives motivate more actors to support the norm. All of these variables played a role in the CBW norm's achieving a cascade. First, the 1874 Brussels Declaration banned use of poisons in war.[19] This declaration was developed by delegates from fifteen key European states at the request of Czar Alexander II of Russia.[20] Once the declaration was drafted, however, some nations were unwilling to accept it as legally binding, so it was not adopted. The significance of the Brussels Declaration's prohibition on poison warfare was that it served as the basis for the much more important Hague conferences. The First Hague Conference (1899) and the Second (1907) were the first efforts culminating in multilateral treaties for the conduct of war.[21] Norm evolution theory identifies "optimal timing" as an environmental factor supporting a norm cascade, and these legal developments and the beginning of an essentially continuous effort to discuss and codify norms for war represent optimal timing for the CBW norm. The First Hague Conference banned the use of "asphyxiating" shells (modern CBW), although no such weapons had yet been invented.[22] The restrictions in the Hague Declaration were intended to apply only to "civilized" states and thus motivated states seeking this status to more deeply embrace this long-standing but now codified CBW norm, thereby seeking legitimacy, reputation, and esteem—consistent with norm evolution theory. This distinction also explains why CW use against Ethiopia by Italy in 1935–36 was technically justified as it was not a conflict between "civilized" nations in the eyes of Europe.[23]

Interestingly, when Price analyzed the proceedings of the Hague Conference of 1899, he found that the basis of the norm—the aversion to poisons—was not the successful argument by norm entrepreneurs and leaders in support of the prohibition on asphyxiating shells.[24] The linkage to poison was not a compelling justification of the need to codify the prohibition as those weapons were already prohibited by custom. Rather, what was compelling was the idea of this new, yet to be invented super poison weapon—modern CW—that could be used "against towns for the destruction of vast numbers of noncombatants, including women and children."[25] It was the idea that CW was a particularly devastating weapon that would affect helpless civilians that led to the success of the norm at the Hague Convention. Notably industrialized CW capable of achieving these feared effects did not yet exist, and it may be that the formal institutionalization of the CBW norm before modern chemical weapons were invented was the single greatest aspect leading to its successful internalization later on. The CBW prohibition in the Hague Declaration of 1899 was particularly important for the norm as it provided the foundation for the Geneva

Protocol of 1925. The Second Hague Conference of 1907 did little to advance or undermine the progress from 1899 and the third Hague Conference was canceled due to World War I.

World War I

While coherence with existing prominent norms is identified by norm evolution theory as a factor leading to successful norm emergence and cascade, it is not without some risk, as evidenced by the complete collapse of the CBW norm during World War I. During the war the CW prohibition in the Hague Declaration broke down on April 22, 1915, in Ypres, Belgium, when German forces used over 150 tons of wind-blown lethal chlorine gas against two French colonial divisions.[26] The Germans justified the use as no more inhumane or cruel than trench warfare, which killed with bullets and howitzers.[27] This contestation of the "humanitarian" core of the norm would again come up in the debates over the Geneva Protocol following the war. The German argument was an attempt to use persuasion to actively undermine the norm and avoid the negative stigma associated with use. However, while the CBW norm was violated, some constraints remained, demonstrating that a norm cascade had been reached despite the major setback of CW use. For example, throughout the war the British, French, Americans, and Germans were all averse to using gas against civilians.[28] In addition to the CBW norm, some argue that this was due to a lack of adequate delivery systems to reach civilians. This development was noteworthy as similar aversions to attacking cities and civilians using airpower eroded during the war. In part, Price argues, this was in response to the Hague Declaration's prohibition and not because CW were perceived as more cruel or inhumane. Further evidence of the limits of the breach of the norm in World War I is the fact that the British restrained their use of CW and used it only in response to German employment (not as first use).[29] However, all sides eventually escalated in their retaliatory use. In some respects the norm actually strengthened through deeper socialization and demonstration as some limits remained and the use of CW against military forces during the war provided many (including future leaders, such as Hitler) with direct and personal exposure to the horror of CW, perhaps validating the innate stigma associated with poisons.

Postwar Institutionalization and Socialization

Despite its prior failure to constrain states, the CBW norm was further embraced after the war. This affirms norm evolution theory's hypothesis that a norm cascade is more likely to occur when institutionalization (through repeated codification) and socialization occur. This socialization and institutionalization sometimes occurred inadvertently as a result of actors and networks pursuing motives contrary to the norm, such as commercial interests or power politics. The next major step in the development of the CBW norm and its cascade was the Washington Naval Conference of 1921–22. This conference was called by U.S. President Warren Harding and attended by nine nations: the United States, Japan, China, France, Britain, Italy, Belgium, Netherlands, and Portugal.[30] During the conference U.S. Secretary of State Charles Evans Hughes succeeded in obtaining a prohibition on CW first use as Article V of the proposed Washington Treaty.[31] This was possible because, as would later be the case with the Geneva Protocol, the participants saw this as simply reaffirming a previous restriction in the Hague Declaration. Participants also saw it as largely unimportant as preparations for using CW would continue (Article

V limited only use, not development), and the terrible experience in World War I left many thinking such restrictions were meaningless anyway once conflict began. Hughes had been prepared to accept a weaker prohibition than was initially proposed, but Article V was accepted as drafted.[32] In pushing for a total ban on first use, Hughes was acting as a strong norm leader and advocate and bucked the conference subcommittee's unanimous recommendation that no special treatment be granted for CW beyond possibly prohibiting their use against cities and other large noncombatant targets.[33] The Washington Treaty did not go into effect, as the French failed to ratify it, but the CW prohibition in Article V went on to serve as the basis for the Geneva Protocol of 1925 a few years later.[34]

In 1925 the Geneva Conference for the Supervision of the International Traffic in Arms was held, and again the United States led the effort to bring codification of the CBW norm to the table. Working with the French, the United States developed a prohibition on use of poison gases, and Poland suggested explicitly adding bacteriological weapons as the distinction between chemical and biological weapons was becoming better understood and both fell under the overarching genre of poison weapons.[35] The final protocol signed on June 17, 1925, said, in part:

> Whereas the *use in war of asphyxiating, poisonous or other gases, and of all analogous liquids, materials or devices, has been justly condemned by the general opinion of the civilized world*; and
>
> Whereas the *prohibition of such use has been declared* in Treaties to which the majority of Powers of the world are Parties; and
>
> To the end *that this prohibition shall be universally accepted as a part of International Law*, binding alike the conscience and the

> practice of nations . . . That the High Contracting Parties, so far as they are not already Parties to Treaties prohibiting such use, accept this prohibition, *agree to extend this prohibition to the use of bacteriological methods of warfare* and agree to be bound as between themselves according to the terms of this declaration.[36]

Thus the Geneva Protocol reasserted the norm-based prohibition on use of chemical weapons and made explicit for the first time that this included biological ("bacteriological") weapons as well—although effective BW had yet to be developed. The protocol was ratified by many, including all the major powers with the exception of Japan and the United States. However, in ratifying the protocol some nations declared it would not be binding if their adversaries violated the protocol. Ironically the U.S. Senate did not ratify the Geneva Protocol despite being the major proponent and norm leader advocating for the prohibition in Geneva as well as in the earlier Washington Treaty. The Senate Foreign Relations Committee recommended ratification in 1926, but Gen. Amos Fries, chief of the U.S. Army Chemical Warfare Service, led an effort to lobby against its ratification by the full Senate.[37] He led a coalition of veterans' groups, chemical manufacturers, and the American Chemical Society and was successful in preventing any U.S. Senate vote on the protocol.[38] Yet although the United States did not formally ratify the treaty, by 1927 it had dismantled its poison gas arsenal and would eventually become a party to the protocol many years later (1975).[39]

While Fries's coalition of interest delayed the United States' ratification of the Geneva Protocol, these same networks were also unwittingly responsible for helping the CBW norm achieve its norm cascade and eventual internalization. In the interwar period (around the debates over the Washington Treaty and the Geneva Protocol), commercial networks and organizations, such as the American Chemical Society that collaborated with Fries, were acting as unwitting norm advocates. The U.S. and British chemical industry lobbied hard to obtain chemical tariffs and investments in military chemical warfare organizations, and in doing so they made "totally irresponsible . . . exaggerations of new weapons developments."[40] These claims regarding the potential effects of CW went largely unchallenged (and were even supported and echoed by opponents of CW, such as Secretary Hughes), which created an unrealistic and disproportionate fear of CW relative to their actual capabilities. Obviously these lobbying efforts backfired on industry and helped establish a foundation for a constraining CBW norm that was further institutionalized in the Washington Treaty and Geneva Protocol.

World War II

The conclusion of the CBW norm cascade stage ended with a bang: World War II. Prior to the war many assumed the norm would at least partially break down as it did during World War I. But CBW were not used on the battlefield between the major powers. Norm evolution theory identifies actor motives of legitimacy, reputation, and esteem as factors that will drive a norm cascade and general adherence, and this experience in World War II is consistent with this hypotheses. For example, while the United States did not ratify the Geneva Protocol, it expressed the general sentiment of the conflict on June 8, 1942, when President Franklin D. Roosevelt said that "use of [CW] has been outlawed by the general opinion of civilized mankind. This country has not used them, and I hope that we never will be compelled to use them. I state categorically that we shall under no circumstances resort to the use of such weapons unless they are first used by our enemies."[41] During World War II many restraints and prohibitions on conflict were violated, but the constraining norm on CBW did not erode insofar as the major powers did not use CBW against each other. Japan did use CBW in its war in Manchuria, much as Italy did against Ethiopia in the interwar period.[42] Part of this restraint was certainly due to a fear of retaliation in kind, but more powerful motivators were the abhorrence of CBW created by the norm and the related public pressure and influence of institutionalization through the Geneva Protocol.[43] Additionally, due to the constraining norm, most of the combatants were not fully prepared for CBW, demonstrating that the norm can combine with other factors in a mutually reinforcing way to achieve nonuse.[44] Perhaps the greatest test of the norm during the war was the Allied invasion at Normandy on D-Day. The German forces had a six-month supply of CW, including 500,000 gas bombs and spray tanks for the Luftwaffe and CW that could have been delivered via artillery. Both sides believed CW use during the invasion could have been decisive as it would have significantly harmed the invading forces and possibly led to the invasion's outright failure.[45] Yet even in this scenario the Germans refrained from first use of CW. A similarly desperate scenario contemplated but fortunately never realized was British use of CW in the event that the Germans invaded Great Britain from the English Channel in 1940. Even in approaching this notional scenario during planning, the CBW norm seemed to hold; British planners dismissed this option as CW use would have raised the question of "whether it really mattered which side won."[46] Another demonstration of the deepening of the norm and achievement of a norm cascade is the fact that the United States also refrained from using CW against entrenched Japanese in the latter stages of the war, despite the fact that they had little reason to fear retaliation in kind (although perhaps another factor was a fear of retribution against prisoners of war).[47] In contrast to World War I, when the norm only partially constrained the various parties, the experience of World War II fully demonstrated that a norm cascade had been achieved and a tipping point reached. Other variables, such as deterrence based on the threat of in-kind retaliation, may have played a role, but the constraining norm was clearly a factor.

Biological and Chemical Weapon Norm Internalization

The third and final stage in the norm life cycle (which may never be reached) is norm internalization. This phase is the effort to codify the norm and achieve universal or near universal adherence.[48] At this stage the norm is so deeply and broadly internalized that it achieves taken-for-granted status. Norm evolution theory outlines some conditions and factors that make norm internalization more likely, specifically the involvement of key legal professions and bureaucracy that institutionalize the norm through legal constructs as well as the existence of professional networks that can socialize individuals to the norm. Norm internalization was reached following the success of the CBW norm during World War II and during the subsequent efforts to build on the Geneva Protocol and develop more robust international agreements and regimes limiting CBW. The exact point when the norm moved beyond the norm cascade stage into the internalization stage is unclear, but these developments in the postwar era clearly represent an effort to deeply codify the norm and achieve universal or near universal adherence.[49] The norm's continued coherence with the deeply rooted poison taboo, perhaps enhanced by regular exposure to harmful chemicals, domestic turmoil, and international arms control bureaucracies, helped support internalization of the CBW norm. However, general secrecy regarding CBW programs and the multiuse nature of the related technology created some challenges and, coupled with the lack of support from key professions for the BW aspect of the norm, impeded deeper internalization. The support of key CW-related professions helps explain why the CW aspect of the norm achieved deeper internalization, consistent with the hypothesis of norm evolution theory.

Factors Supporting Internalization

One explanation for the increasing relative strength and internalization of the CBW norm in the latter half of the twentieth century is regular personal exposure to harmful chemicals coupled with coherence with the deeply rooted poison taboo. These coherence and "demonstration" factors are identified by norm evolution theory as important to achieving a norm cascade. While most people will remain blissfully unaware of the effects of chemical or biological weapons, let alone be exposed to them, people are regularly confronted with harmful chemicals and substances in the form of pesticides, pollution, or cleaning solvents. Direct experience with or secondhand stories of harmful accidents can help bolster the basic aversion to poison and CBW and thus indirectly strengthen the constraining norm.[50] This deepening of the norm after its success in World War II led to efforts to internalize and codify not only its constraints on use (no first use, no attacks on innocent combatants, and so on) but to move deeper in constraining development and proliferation of these weapons and ultimately seek total disarmament. This is similar to the path followed by the nuclear norm during this same period. This is only an indirect explanation that perhaps helped create more fruitful soil for internalization; it took other factors to really achieve internalization.

Domestic Turmoil

Norm evolution theory identifies domestic turmoil as a key motivating factor that makes states and key actors more likely to support and participate in norm internalization. The CBW norm's experience is consistent with this hypothesis. For example, the first major step in this new direction of internalizing the CBW norm was specific to BW. In 1969 the United States renounced biological weapons, destroyed its stockpile, and terminated its twenty-seven-year-old offensive BW program. The United States took this bold step due to a variety of factors, all influenced by the CBW norm. The realist calculus that the country was highly vulnerable to BW attacks and that BW would be widely accessible to smaller, weaker states certainly played a role.[51] However, as with earlier developments with CW, this cost-benefit calculus was influenced by the norm in the sense that the norm had prevented BW from playing a major role in past conflicts and military planning in general (although limits in technology and their delayed and unpredictable effects also played a role).[52] The United States decided to forgo these weapons and further delegitimize them in the hopes of dissuading other states from developing them. Additional actor motives were consistent with norm evolution theory, which hypothesizes that actors are more likely to support norm internalization when motivated by conformity and domestic turmoil. Domestic turmoil in the United States over the role of the military and the conflict in Vietnam may have helped motivate President Nixon to disavow BW.[53] Harvard University biologist Matthew Meselson was a lead norm entrepreneur in this effort and started advocating against CBW and thus indirectly for the constraining norm in the Johnson administration and then in the Nixon administration.[54] Nixon's renunciation was truly historic as it was the first time in history that a major power unilaterally abandoned an entire weapon category.[55] In addition to disavowing BW, the United States also declared a policy of no first use for lethal chemicals.

Lack of Support from Key Professions

Following the United States' unilateral disarmament, the BWC, the first treaty to outlaw an entire class of weapons, was negotiated and opened for signature in 1972. The BWC declared the use of BW to be "repugnant to the conscience of mankind" and built on the Geneva Protocol, which prohibited the development, production, and stockpiling of BW by any of the now 155 states parties.[56] Norm evolution theory identifies the participation and support of key professions as important to norm internalization. The internalization of the BW aspect of the CBW norm lacked this support, and, as predicted, internalization suffered accordingly. For example, while hailed as the first multilateral treaty to ban an entire category of weapons, the short, fifteen-article BWC includes no verification mechanism.[57] The lack of a verification mechanism was due in part to the lack of interest and incentive by U.S. negotiators in demanding such a mechanism as the United States had recently unilaterally renounced its own BW program.[58] Further, when the BWC was adopted in the early 1970s, there was a view that there were no reliable technical means to verify that nations were not researching, developing, or stockpiling biological weapons.[59] Perhaps most important, the wide range of actors and professions associated with the industrial sectors impacted by a potential BWC verification regime made their concerns known. This includes the biotechnology, pharmaceutical, dairy, brewing, and industrial baking industries, which use technologies and materials to develop and manufacture legitimate products such as medical therapeutics that could potentially be diverted to make biological weapons with minimal modifications.[60] Although the treaty lacked verification measures, its primary purpose was to delegitimize BW and stigmatize them by reinforcing the norm against their use (as was done with the Geneva Protocol almost fifty years earlier) and also their development and possession.

Secrecy and Multiuse Technologies

Norm evolution theory does not directly address the issue of secrecy or multiuse technologies, but they have clearly played a role in these challenges with deeper internalization of the CBW norm. As the BWC significantly expanded the CBW norm's internalization into development, proliferation, and disarmament, it dealt with some new challenges—some associated with the verification challenge—that did not face the more basic institutionalization of the usage norm. The BWC is a "general purpose criterion" treaty. This means that it exempts peaceful activities that could be associated with biological or chemical weapons unless such activities are intended for hostile purposes.[61] Essentially it is the purpose for which the biological components were intended that determines if activities involving the substances would be prohibited under the treaty. This "purpose-based" approach is necessary due to the multiuse nature of both chemical and biological activities. Without it, any such treaty would dramatically limit the peaceful application of related technology and would be totally unfeasible. Moving beyond the "dual-use" term applied to WMD technology that also has a peaceful application, Gregory Koblentz proposes the term *multiuse*, which more appropriately describes the dilemma whereby chemical or biological technologies have a multitude of peaceful applications in addition to their weaponization potential.[62] For example, botulinum toxin is one of the most lethal substances on the planet and can be a very potent BW agent, but in extremely diluted form it is used to treat muscle spasms and combat wrinkles through cosmetic Botox treatments. There is the added complication that defensive efforts intended to protect against BW involve activities very similar to an offensive program.[63] For example, in order to defend against biological agents, effective medical countermeasures such as vaccines are developed, and the pathogenic agent often must also often be produced or utilized. Therefore the BWC sought to codify the CBW norm while allowing these legitimate activities as long as they are of a peaceful or defensive nature. Norm evolution theory indicates that internalization is more likely to occur when key factors such as professions and bureaucracies are involved. Clearly the lack of enthusiastic involvement of key biological industries limited the norms internalization in regard to institutionalizing a verification mechanism.

The failure of this full internalization of the BW norm aspect related to BW development, proliferation, and disarmament was not without consequences. Even after the Soviet Union signed the BWC and publicly disavowed BW, they were secretly ramping up their BW program. The lack of such a BWC verification mechanism became a widely acknowledged deficiency following the 1979 anthrax event at Sverdlovsk (now Yekaterinburg) in the Soviet Union, when the United States and other members to the convention lacked a clear mechanism to challenge widely suspected Soviet violations of the BWC specifically and the CBW norm generally.[64] The Soviet program ultimately became the largest biological weapons program in history, and the Soviet Union managed to keep this violation of the BWC secret for decades.[65] However, in spite of this obvious failure of norms and international agreements to fully constrain some BW programs, the norm continued to prevent use during the conflicts that did occur and likely constrained proliferation and development by forcing it to occur in the shadows.

Arms Control Bureaucracies

In part due to the failure of the BWC and the CBW norm it institutionalizes to prevent the illicit Soviet BW program, norm leaders and the arms control bureaucracy have continued efforts to negotiate a legally binding verification mechanism for the treaty. Norm evolution theory identifies these international bureaucracies, law, and lawyers as key actors that make norm internalization more likely. This is what has occurred with the CBW norm. Since the third BWC Review Conference (REVCON) in 1991, BWC member states have formally pursued a mandatory verification protocol. At this REVCON the member states tasked a group of experts to explore these verification measures, which culminated in a report to a special meeting of states parties in 1994. Following this report member states established another group, known as the Ad Hoc Group, to build on the work of the group of experts and develop a viable compliance protocol to be considered at the 2001 REVCON.[66] Over a six-year period the Ad Hoc Group met to address the issue of a verification mechanism for the BWC. Very quickly a divide became apparent. Western and developed states were interested in negotiating a verification protocol to address the threat of biological weapons, while states associated with the Non-Aligned Movement were, in the view of some, focused on a verification protocol that would allow or facilitate the transfer of advanced biological technology from developed to developing states.[67] The Ad Hoc Group's chairman, Tibor Tóth, attempted to put forward a compromise draft verification protocol in March 2001 to address some of these disagreements.[68] In July, only a few months after it was originally proposed, the United States rejected the protocol because it concluded that the verification regime would have given other nations too much visibility into sensitive biodefense efforts as well as exposed the chemical industry's important proprietary biological information, all while weakening export regimes that are helpful in achieving real nonproliferation goals.[69] Once the United States came out against the BWC verification protocol, the effort quickly unraveled and has since been unable to move forward. These more recent experiences with internalization again affirm norm evolution theory's hypothesis that norm internalization is more likely to occur if international bureaucracies, law, and lawyers (working with and as diplomats to increasingly legalize warfare) as key actors are involved and supporting the norm. In addition, nonnormative factors such as the lack of opposition by great powers helped contribute to its success. Further, the lack of industry and bureaucratic support from the United States and some other Western countries continued to block deeper internalization through a BWC verification mechanism.

Support from Key Professions

While the legal codification of the development, proliferation, and disarmament aspects of the constraining CBW norm first occurred for BW, it did eventually occur for CW as well. The CWC, which entered into force in 1997, was negotiated over twenty years and similarly prohibits the development, production, stockpiling, and use of an entire category of weapons (in this case CW). However, unlike the BWC, it includes a verification mechanism to ensure compliance with the treaty's obligations, in part thanks to the critical involvement of the chemical industry (a key profession as identified by norm evolution theory).[70] The CWC verifies this peaceful intent of chemical facilities as well as mandated chemical weapon disarmament efforts through the Organization for the Prohibition of Chemical Weapons (OPCW) and its Technical Secretariat.[71] The OPCW requires states parties to file annual declarations, and specially trained OPCW inspectors conduct a variety of verification inspections.[72] As of early 2014 the OPCW had conducted 2,599 inspections at roughly 1,940 military and industrial sites in eighty nations.[73] As of October 2013 there were 190 states parties, and only four states had not signed or acceded to the convention.[74] This new bureaucracy, created by the CWC, significantly enhances the internalization of the CBW norm as it creates the actor mechanism of habit and deepens its roots in a broad professional network. Norm evolution theory indicates that both of these are key elements in norm internalization.

In contrast with the BWC, the stronger internalization of the CW component of the CBW norm was due to the successful support of key professions. It is worth examining how this occurred in order to garner lessons for future cultivation of key professions for norm evolution and internalization. The chemical industry was first invited to the CWC negotiating table by Ambassador Adrian Fisher of the U.S. Arms Control and Disarmament Agency in December 1978, when he requested a meeting with industry representatives to obtain their input and assistance.[75] From this point forward the chemical industry (principally represented by the Chemical Manufacturers Association, known at the time as the CMA and now known as the American Chemistry Council) became increasingly involved in the U.S. efforts to negotiate the CWC. Will Carpenter (who was a CMA participant in the process) and Michael Moodie identify two main reasons why the chemical industry responded favorably to this invitation to become involved in the CWC effort. First, the industry had learned that in general key issues associated with government efforts such as the CWC needed to be identified extremely early in the process before they got too much attention from the public, media, and government in order for industry to be able to shape the outcome and play a positive role. Second, industry determined that it needed to help government develop its objectives and strategies because the less informed the government was, the more likely its efforts would be deficient (and would conflict with industry objectives and strategies). Carpenter and Moodie also identify three factors that led the chemical industry to conclude that the CWC was inevitably going to be a significant issue moving forward and consequently they should become involved. First, according to the CMA, the CWC was unavoidably going to become an issue for them because they were already heavily regulated and it was likely regulations would be included in the convention. Second, momentum for WMD arms control agreements was building. Third, verification was likely to be included in the CWC since the lack of verification in the BWC was widely acknowledged as a problem and any new treaty would be designed to avoid this problem.[76] The establishment of a dialogue between the industry and government following their first interaction on the CWC in 1978 ultimately convinced the CMA that it had an opportunity to successfully and proactively engage on the issue.

Industry involvement helped give negotiators the additional technical information they needed regarding the costs and benefits (for governments and industry) of various declarations and inspection requirements. This in turn gave negotiators additional confidence and insight that helped lead to negotiating the CWC. It is possible and perhaps likely that the convention would not have been agreed to had such technical confidence been lacking. Without such extensive industry involvement the chemical industry may have likewise lacked confidence in the outcome and actively opposed CWC adoption (as it did the Geneva Protocol decades earlier, delaying its ratification), further jeopardizing the CWC's adoption by the negotiating governments. The chemical industry, perhaps due to its active role in developing the convention, embraced its implementation. On the tenth anniversary of the CWC's entry into force, the International Council of Chemical Associations reflected this broad industry support of the CWC when it issued a statement saying that the CWC "has been a model of a vigorous regulatory system that ensures sensitive chemicals are not diverted to illegal uses, without an adverse impact on legitimate commercial activities. It was the first multilateral disarmament treaty that considered the essential chemical industry, and we recognize the decade of progress that the treaty has yielded."[77] The chemical industry has remained involved in efforts to fine-tune the CWC at the three REVCONs held since its enactment as well as at other OPCW-sponsored meetings.[78]

The Conformity Motive

Norm evolution theory identifies conformity as a major motive driving norm internalization. For the CBW norm, conformity played a major role in furthering the norm's internalization even when violated as it forced violators to attempt to hide their violations and led to public and widespread condemnation when violations were recognized. While the CWC was being negotiated and not long after the BWC went into force, the second greatest example (eclipsed by only World War I) of the failure of the CBW norm to successfully constrain an actor occurred: The Iran-Iraq War (also known as the First Persian Gulf War). This conflict began in September 1980, when Iraq invaded Iran following a series of border disputes, and did not end until August 1988 with a cease-fire under the auspices of United Nations Security Council Resolution 598.[79] During the war, as Iran began to make progress against Iraq and repelled the invasion, Iraq resorted to CWs. The CIA estimated that Iran suffered over fifty thousand casualties from these CW attacks, not including collateral civilian casualties.[80] This attack was condemned by the UN Security Council in a March 21, 1986, declaration stating that "members are profoundly concerned by the unanimous conclusion of the specialists that chemical weapons on many occasions have been used by Iraqi forces against Iranian troops, and the members of the Council strongly condemn this continued use of chemical weapons in clear violation of the Geneva Protocol of 1925, which prohibits the use in war of chemical weapons."[81] The United States was the only vote against the statement due to its support for the Iraqi regime and its ongoing poor relationship with Iran. In this case, while United States continued to press for conformity with the norm, its role as a norm leader was secondary to its interest in ensuring Iraq did not lose the conflict. This is a good illustration of the multiple variables that influence state behavior and how at times material interest can trump the influence of norms. However, U.S. officials did indicate that they would not have tolerated the use of CW on anything other than military objectives, similar to how the most basic aspect of the CBW norm, prohibiting use on civilian targets, was sustained during World War I.[82] Further evidence that CW use was not a total collapse of the CBW norm and that the motive of conformity was a potent force in keeping the norm alive was the fact that Iraq did not acknowledge using the weapons until the end of the war, and even then claimed that they supported the CW prohibitions but had to use them in order to defend their homeland.[83] After the war ended, President Reagan addressed the UN General Assembly in September 1988 and called for a conference to review the "rapid deterioration of respect for international norms against chemical weapon use." The conference was held in Paris the following year and resulted in the declaration that participating nations "solemnly affirm their commitments not to use chemical weapons and condemn such use" and also recognized the importance of the Geneva Protocol of 1925 and called on nations that had not yet acceded to the protocol to do so.[84] These actions bolstered the international pressure to conform to the norm, despite the recent violations.

Despite this evidence that the CBW norm was still present, although not effective in fully constraining Iraq, the use of CW coupled with no tangible consequences for Iraq fueled some broader backsliding of the norm internalization, and "more countries than ever began to develop [CW]."[85] This negative development demonstrates that even when norm internalization occurs, various factors can permit a norm to backslide. A norm is not strengthened when it is violated without serious consequences as it makes the norm appear weak. In the 1990s and early 2000s the international community seemed to learn from the negative consequences of their weak response to Iraqi CW use and were more firm when addressing potential violations of the CBW norm. The norm was fully sustained during the Persian Gulf War in 1991, and Saddam Hussein did not use CW. The CIA attributes this nonuse to the fact that the United States and other coalition members repeatedly made stern warnings that dire consequences would result if the norm was violated and CW were used, illustrating that sometimes material incentives are used to bolster adherence to a nonmaterial norm.[86] A senior Bush administration official agreed that the decision not to employ these weapons seems to have been based on the understanding that initiating chemical warfare would cross a "red line beyond which all previous bets are off."[87] Clear and specific consequences were again linked to the norm—additional factors that norm evolution theory indicates are important. Following this conflict, the international community continued to insist on consequences for failure to comply with the norm. For example, in April 1998, when UN inspectors determined that Iraq still had not issued the required "full, final, and complete disclosure" about its illicit CBW programs, the UN continued to demand full compliance before lifting crippling sanctions. This insistence was part of President Clinton's view that "failure to obtain full accountability from Iraq would signal to others that they can develop biological and chemical weapons with impunity. The consequence would likely be a twenty-first century rife with these weapons."[88]

Following the September 11, 2001, terrorist attacks, the global resolve to address CBW threats (characterized as WMD, along with nuclear threats) was only strengthened, leading to the controversial decision to invade Iraq in 2003. Regardless of the negative consequences of that decision and subsequent occupation, the decision to use material force and invade only furthered the internalization of the CBW norm and conformity motives by demonstrating the resolve of key actors to address potential deviation from the norm. However, this may have had counterproductive effects by incentivizing states such as Iran and North Korea to accelerate rather than decelerate their WMD programs so as to obtain the weapons and thus deter a norm-motivated attack. One positive effect of the invasion of Iraq was to help convince Muammar Gaddafi to renounce Libya's WMD programs and welcome international inspectors in 2003.[89] In 2013 the world was faced with another violation of the norm, this time in the context of Syria's ongoing civil war. Ultimately UN weapons inspectors confirmed that the binary nerve agent sarin was used in an attack in a suburb of Damascus on the morning of August 21, 2013.[90] Unlike during the 1980s, the international community reacted more strongly, with the United States indicating that this CW attack crossed a "red line" and threatening serious consequences, potentially including a military strike.[91] While there was disagreement over a military versus diplomatic response, the outcry was unanimous. Ultimately this robust response, a sharp contrast to the international response to Iraq's CW use against Iran, led to Syria acceding to the CWC and OPCW inspectors beginning to dismantle Syria's illicit CW program, although in early 2014 the regime was accused of dragging its feet.[92] These developments indicate that the backsliding in norm internalization that occurred in the 1980s was temporary and that internalization of the CBW norm continues. Full internalization (taken-for-granted status and near universal or universal compliance and codification) has not yet occurred, but such a state may never occur for a norm.

Summary

While the relative impact of the CBW norm varied, and certainly other variables influenced state behavior, its emergence, cascade, and internalization of constraining norms for chemical and biological weapons was consistent with many of the various hypotheses from norm evolution theory. The emergence of modern CBW norms was extremely rapid, rising from a perhaps biologically rooted taboo against the use of poison weapons. As a result, norm entrepreneurs and organizational platforms (sometimes self-motivated) were readily available to support norm emergence and exploit the optimal timing of international organizations, and the legalization of war using important mechanisms and factors such as framing, prominence, and coherence helped the modern CBW norm emerge from this more primitive basis in the late 1800s. The opportunistic inclusion of the total ban on poison weapons in the 1899 Hague Declaration—before such weapons even existed—proved important in the eventual internalization of the norm, although it would collapse during World War I and have to reemerge. This preemptive prohibition was unique in that a new technology of warfare was anticipated as a "preemptive proscription" put in place, giving the norm more universal force once the weapons did exist.[93] Thus the CBW norm was established internationally and then diffused.[94] The Hague Declaration created a distinct category of weapons—gas shells—that granted a special "unconventional" status to CBW, although coherence with humanitarian norms regarding noncombatant immunity was perhaps more important in its initial spread. During World War I this weaker humanitarian argument came under direct assault as the Germans explicitly argued that CW was no less humane than the guns and howitzers, which made life in the trenches such a "terrible hell." Similar arguments were made in the U.S. Senate and led to the delay in ratification of the Geneva Protocol of 1925. However, constraining CBW norms did not rely solely on this humanitarian argument and could fall back on their "unconventional" characterization and special status and linkage to the poison taboo. This helps explain why CBW were not used in many regional wars where the adversary possessed no ability to respond in kind, such as the Spanish Civil War, the Korean War, the French war in Algeria and Indochina, the Vietnam War, and (allegedly) the Soviet war in Afghanistan.[95]

From the Hague Declaration in 1899 to World War II, the norm experienced some significant setbacks but ultimately achieved a norm cascade (in part due to contingency and chance interacting with other variables), and deeper internalization beyond simply constraints on CW and BW use began. The institutionalized CBW norm may have developed, at least in part, to complement the interests of the major powers at the beginning of the twentieth century and between the United States and Soviets during the Cold War, yet as the memories of those interests faded the norm remained and deepened as an independent force.[96] As internalization continued in the areas of development, proliferation, and disarmament, the primordial foundation associated with poisons was evident as a special sense of repugnance and abhorrence was reserved for BW and CW that were not mentioned in any other arms control agreements or treaties; only the Geneva Protocol of 1925, BWC, and CWC included such language.[97] The association of CW and BW with poison has been important in sustaining and institutionalizing the constraining CBW norm in the face of some setbacks.[98] Additionally, key political leaders played a critical role in cultivating the moral revulsion underlying the norm at just the right moments. For example, at the conference developing the Washington Treaty, the British delegate Lord Balfour forcefully and decisively argued that "agreement on the prohibition against CW was desirable in order to do something to bring home to the consciences of mankind that poison gas was not a form of warfare which civilized nations could tolerate."[99] As another example, Nixon led the effort to negotiate the BWC by unilaterally abandoning the U.S. offensive BW program. Moving into the second decade of the twenty-first century, the norm is increasingly (although not fully, a case in point being Syria in 2013) internalized and remains a potent force of restraint in international politics and war. Table 5 summarizes

Table 5. Hypotheses of norm evolution theory applied to CBW

HYPOTHESES	CBW
NORM EMERGENCE	
Significant presence of norm entrepreneurs/norm leaders with organizational platforms	X
Key actors motivated altruism, empathy, and ideational commitment	X
Actors employ persuasion, framing, memes, shame	X

Table 5. (*continued*)

HYPOTHESES	CBW
Domestic political concerns	X
State distribution of power, prevailing levels of technology, availability of natural and human resources, etc. favoring the norm	X
Large-scale turnover of decision makers	
Recognized failure of norms from a previous norm generation	
Emergence of a new issue area where prevailing norms are not well established	
Norm prominence	X
Norm coherence—connections to key ideas, adjacency, and grafting	XX
NORM CASCADE	
Key actors are involved, specifically states, international organizations, and transnational networks	X
Key actors motivated by legitimacy, reputation, esteem	XX
Actors employ persuasion and diplomacy	X
Actors employ institutionalization	XX
Actors employ socialization	XX
Actors employ demonstration	X
Optimal timing	XX
Norm clear and specific	X
Norm makes universal claims*	X
Norm coherence	XX
NORM INTERNALIZATION	
Key actors are involved, specifically law and lawyers, professions, and bureaucracy	XX
Actors motivated by conformity or domestic turmoil	XX
Actors employ habit and institutionalization	XX
Existence of professional networks	XX
Norm clear and specific	X
Norm makes universal claims*	X
Norm coherence	XX

* Universal claims may still be selective in the sense that they apply only to "civilized" nations.

More prevalent variables are indicated by XX rather than X.

the hypotheses of norm evolution theory that were present in the history of the evolution of norms for CBW.

While overall the hypotheses of norm evolution theory are consistent with the experience of the constraining CBW norm's emergence, cascade, and internalization, the CBW norm experience offers some additional insights to complement the general hypothesis of norm evolution theory. The impediment and challenge that multiuse technology offered to norm evolution may be applicable to other emerging weapons that share this trait. Additionally, the experience highlights the unintentional role certain actors, such as key professions and bureaucratic advocates of the particular weapon, can play in impacting the emerging norm's growth. The internalization of aspects of a norm governing usage occurs more rapidly and is easier to achieve than aspects governing development, proliferation, and disarmament. These lessons provide a foundation for the development of a refined norm evolution theory for emerging-technology weapons in chapter 5, which will then be used to offer predictions on the evolution of norms for emerging-technology cyber weapons in chapter 6.

3 Norm Evolution for Strategic Bombing

> I will not wage war against women and children! I have instructed my air force to limit their attacks to military objectives. However, if the enemy should conclude from this that he might get away with waging war in a different manner he will receive an answer that he'll be knocked out of his wits!
>
> Adolf Hitler

As with cyber weapons today, with the advent of aircraft (initially balloons and airships, followed by airplanes) and their ability to drop munitions from the air during conflict, states had to grapple with a brand-new technology and approach to warfare. One method of using this new airpower technology became known as strategic bombing, which is the employment of airpower—often as part of a total war—against nonmilitary targets (civilians and economic infrastructure) in order to destroy a nation's war-fighting capacity or break enemy morale. For this reason it is also known as "terror bombing" or "morale bombing."[1] Similar to chemical and biological weapons, strategic bombing shares some characteristics with cyber warfare. For example, strategic bombing made civilian populations highly vulnerable, was difficult to defend against, and used technology that also had peaceful applications (air travel and transport)—all of which can also be said about cyber warfare today. Accordingly, examining norm evolution for strategic bombing is another valuable case study to help extend norm evolution theory and develop expectations for how norms for cyber warfare will develop. This chapter examines the norm life cycle for strategic bombing and the various hypotheses from norm evolution theory. Overall the hypotheses of norm evolution theory are consistent with the experience of the constraining strategic bombing norm's emergence, cascade, and internalization.

Airpower and strategic bombing were totally novel in warfare. The idea of bombing from the air preceded the capability and began the process of norm emergence. In his history of strategic bombing, Lee Kennett identifies the fact that "the bomber was an idea long before it was a reality" as a major discrepancy in airpower's history.[2] Hot air balloons originated in China in the third century, and Gen. Zhuge Liang used balloons called "Kongming lanterns" to signal to his forces and frighten enemies.[3] In the Western world the first hot air balloon flight occurred near Paris in November 1783, and in 1794 the French revolutionary government created a corps of balloonists (known as *aerostiers*).[4] The aerostiers were focused on observation and reconnaissance and not delivery of ordinance, but it wasn't long before arming them was considered. J. C. G. Hayne, a Prussian lieutenant, soon published a book on "military aeronautics" that theorized that balloons could be made to carry cannons or drop "grenades and other harmful things."[5] Schemes to use balloons to drop explosives were proposed or pursued by the Russians against Napoleon's invading forces in 1812, during the Mexican War in 1846–48, and against Venice during the Austrian Hapsburg revolt.[6] From the earliest days, aerial bombing was viewed as a tool for achieving kinetic destruction and one that could achieve "psychological shock and social destruction . . . [that] would impair both the enemy's capacity and will to fight."[7] Beyond their use as elevated observation posts, however, the actual military utility of airships was very limited during this early time, capable of achieving neither the physical or psychological impact envisioned. This led some to be rather skeptical about the future military utility of airpower. For example, as late as 1908 the U.S. astronomer William H. Pickering stated, "Another popular fallacy is to suppose that flying machines could be used to drop dynamite on an enemy in a time of war."[8] Some saw aircraft in an entirely different light, as eventually leading to the end of international conflict not due to its horrific military potential but because, in the words of Victor Hugo in 1862, it would bring about a "peaceful revolution" by achieving the universal abolition of borders.[9] Over the past century these preliminary notions of airpower and strategic bombing were put to the test with horrifying results. A constraining norm for strategic bombing—based on norms associated with noncombatant immunity—began to emerge even before heavier-than-air aircraft had been invented. This norm would falter and then ultimately collapse during World War II. Yet in the postwar era the constraining strategic bombing norm reemerged to achieve a norm cascade and ultimately become internalized by much of the international community. How did this happen? Overall the hypotheses of norm evolution theory are consistent with the experience of the constraining strategic bombing norm's emergence, cascade, and internalization. Unlike the norms for chemical and biological weapons, regulative norms for strategic bombing arose largely regarding use and not for the development, possession, and proliferation of bombers. Norm evolution theory successfully predicts many of the factors that were important in this process, including coherence and grafting onto norms for noncombatant immunity; the distribution of power and technology; altruism, empathy, shame, and conformity motives; a growth of organizational platforms and international bureaucracies; the recognized failure of norms; and optimal timing. The strategic bombing norm experience offers some additional insights to complement the general hypothesis of norm evolution theory, including the risks of inadvertent violations due to new issue areas and technology weakening a norm and the ability of real-time compliance visibility through global media to facilitate internalization. In this chapter I test norm evolution theory hypotheses against the historical record for norm development for strategic bombing, specifically regarding the actors, actor motives, and important mechanisms and factors supporting norm emergence and growth during each stage of the norm life cycle.

Strategic Bombing Norm Emergence

As discussed previously, norm emergence is the stage wherein a new norm initially comes into existence, often tenuously. Norm evolution theory highlights several specific conditions that make norm emergence more likely: the number and type of actors involved, important mechanisms these actors can employ to promote the norm (such as persuasion and framing), and various environmental and norm factors that can further support norm emergence (distribution of power, prevailing levels of technology, norm coherence, and so on). The emergence of constraining norms for strategic bombing was consistent with these various hypotheses from norm evolution theory, and coherence with the notion of noncombatant immunity played a key role in the norm's initial emergence. The strategic bombing norm emerged quickly, even before the technical capacity to conduct strategic bombing existed, and early norm entrepreneurs were motivated by self-interest as much as empathy and altruism. However, factors such as the lack of clarity regarding the application of the norm, speculation that strategic bombing would shorten war and was therefore more humanitarian, and the generally unproven nature of the technology led to miscalculation and faltering during World War I and total collapse during World War II.

Coherence and Grafting

The strategic bombing norm initially developed largely as a result of its coherence with and grafting onto existing norms. While aircraft were a totally new invention, norms regarding their possession and use did not emerge out of the ether. They were the product of the existing norms constraining armed conflict, particularly the norm of noncombatant immunity. This linkage would be problematic later on, yet it effectively served as the preliminary catalyst for the strategic bombing norm.[10] This concept traces its roots back to the feudal era and the Middle Ages, when the theologian Thomas Aquinas articulated the concept that war requires a "complete prohibition on killing the innocent."[11] In the modern era the idea of noncombatant immunity was codified in the Lieber Code, named after its author, Francis Lieber, and formally called the *Instructions for the Government of Armies of the United States in the Field, General Order No. 100*, which provided comprehensive regulations for the conduct of land warfare.[12] Toward the end of the century this concept provided the lens through which strategic bombing would come to be viewed.

In addition to dealing with chemical weapons, the Hague Conference of 1899 also dealt with airpower.[13] Russia was the strategic bombing norm leader at this conference, motivated in part by its recent failure to develop a dirigible airship while the French and Germans were succeeding.[14] Britain opposed the proposed prohibition, seeing balloons as a potential tool to offset their army's numerical disadvantage when compared to France and Germany.[15] The Americans were the ones who brokered the compromise between the British and Russian position; U.S. delegate William Crozier suggested a five-year term for the constraints on aerial bombing. The U.S. support for this term was also based on the fact that technology at the time made such bombing ineffective and indiscriminate and that it might actually become more effective and help end wars sooner. Ultimately the conference declaration placed a "five-year moratorium on the launching of projectiles and/or explosives from the air."[16] The five-year limit on the prohibition ultimately made it much less meaningful.[17] Another aspect of the conference dealt largely with conduct for war on land but reached agreement on issues relevant to airpower. Article 25 of the conference declaration stated that "the attack or bombardment of towns, villages, habitations, or buildings which are not defended is prohibited"; Articles 26 and 27 required an attacking force to warn authorities before a bombardment began and avoid striking nonmilitary targets.[18] These latter elements were not explicitly linked to aerial bombardment, but that was not necessary as any aerial bombing was prohibited by the five-year moratorium. Unfortunately the declaration did not define the term *defended*; this would lead to a great deal of confusion and misunderstanding later, particularly after the Second Hague Conference. According to airpower historian Tami Davis Biddle, the environment at the conference was cautious and delegates did not want to give an advantage to their adversaries, so they did not extensively and publicly offer their various interpretations of the term.[19] A German delegate, Colonel Gross von Schwarzhoff, did assert that the norm expressed in Article

25 did not prohibit the destruction of any buildings whatsoever, and no other participant objected to this interpretation.[20]

Self-Interest Motive and Distribution of Power and Technology

In addition to coherence and grafting, norm evolution theory identifies the state distribution of power and technology as key environmental factors that can facilitate norm emergence. These factors were at play with the early emergence of the strategic bombing norm as states, often motivated by self-interest, grappled with the norm in various international venues. The Second Hague Conference, held in 1907, again addressed the issue of strategic bombing, among other things. Since the previous conference a great deal had happened. Zeppelins, invented by the German Count Ferdinand von Zeppelin, were rigid airships equipped with a light-alloy skeleton that allowed them to be much larger than a regular hot air balloon.[21] Of course, this dramatic increase in size meant that there was more room for payload, potentially including explosives and bombs. In late 1903 the Wright brothers had conducted the first powered and sustained heavier-than-air human flight. While initially appearing less capable for bombing purposes, airplanes added a whole other dimension to the prospect of strategic bombing. In this context, the delegates at the Hague Conference in 1907 brought up the renewal of the prohibition on aerial bombing, which had expired three years earlier. The position taken by the delegates was largely dictated by their own interests based on their nation's relative air strength.[22] The states with robust and growing aerial programs (Germany, Italy, and France) opposed continuing the clear prohibition on aerial bombing. The Italians in particular did not feel threatened by aerial bombardment as the dirigibles of the day were unable to fly over the Alps or the Dolomites and could be easily intercepted if coming over sea.[23] Britain, lacking a significant air force and fearing a threat to its naval superiority, supported renewing the clear codification of the emerging norm against strategic bombing.[24] Austria-Hungary, Greece, Portugal, China, and Turkey all joined the British in supporting the norm.[25] This divergence of interests among prospective norm entrepreneurs, coupled with the fact that the previous codification of the constraining strategic bombing norm had already expired and thus the status quo was not a codified norm, resulted in little progress for the norm. The clear prohibition on aerial bombing was not renewed. Kennett described this decision not to fully prohibit strategic bombing while the technology was still embryonic as virtually ensuring that the practice would spread and intensify.[26] The Land Warfare Convention from the First Hague Conference was amended (specifically Article 25) to make it clearly inclusive of aerial bombing; it now stated, "The attack or bombardment, *by whatever means*, of towns, villages, dwellings or buildings which are undefended is prohibited" (emphasis added).[27] However, the term *undefended* was still not defined, further evidence that the emerging strategic bombing norm arose and evolved from existing norms constraining armed conflict, particularly the norm of noncombatant immunity. This concept from the Middle Ages of undefended towns (occupied by noncombatants) was intended to prohibit an attack on places that could be occupied without military force.[28] An undefended town was understood as a town "open" to occupation by not being fortified and lacking an active military presence.[29] However, this distinction was much murkier in an age of increasing industrialization and military mobilization and defenses and fortifications around cities less relevant to modern warfare. The legal scholar Percy Boardwell offered a leading interpretation of the prohibition on bombardment of undefended towns after the conclusion of the Second Hague Conference: "Only where there are no fortifications, no troops, and no open resistance by the population, does this article apply."[30] His interpretation was not universal, as evidenced by British debates on whether placing antiaircraft guns in London would make it a defended city.[31]

Another outcome of the Hague Conference of 1907 relevant to the strategic bombing norm was its declaration on naval bombardment during war, referred to as the Naval Convention. Article 2 of the Naval Convention identified specific targets that were to be considered legitimate military targets, such as "depots of arms or war material, workshops or plants which could be utilized for the needs of the hostile fleet or army." This helped clarify that industrial targets related to the war effort were legitimate targets of bombardment. Additional elements of Article 2 also specified that commanders were not responsible for "unavoidable damage which may be caused by a bombardment," essentially acknowledging that collateral damage is an unavoidable aspect of bombardment.[32] However, this additional information spawned some confusion about the emerging norm as it appeared to sanction greater flexibility in bombardment for naval forces compared to land forces.

Ultimately the emerging constraining norm for strategic bombing did not fare well at the Second Hague Conference; the relevant portions of the Land Warfare Convention and Naval Convention were ambiguous at best and clearly permitted aerial bombardment of certain cities and targets. This was a significant step backward compared to the total (albeit temporary) constraining norm codified at the First Hague Conference and was the result of the allure of new technology and divergent state interests. This is consistent with norm evolution theory, which indicates that the environmental factor of state distribution of power and prevailing levels of technology (in this case, national aerial military programs) play a major role in how rapidly a norm emerges. In this case the balance of power and general distrust in legal prohibitions to constrain adversaries made potential norm entrepreneurs and leaders less willing to promote the new norm.

Public Awareness, Altruism, and Empathy

Norm evolution theory suggests that altruism and empathy (in this case toward potential noncombatant victims of strategic bombing) are often motivating factors for norm entrepreneurs and leaders. This can lead to a positive norm environment by creating domestic political pressure for states to adopt the norm (as was seen with the emerging nuclear norm). While public fascination with airpower existed at this time, there had not yet been a real demonstration of airpower, and it was quite reasonable to believe it would never move beyond its current role as a signaling and scouting resource to assist the ground forces. The genre of science fiction was coming of age with authors like Jules Verne and H. G. Wells. Verne wrote a book in 1887 titled *Robur the Conqueror*, also known as *Clipper of the Clouds*, which captivated readers with its story of a giant electric heavier-than-air airship called the *Albatross*. Verne's sequel, *Master of the World*, published in 1904, introduced another aircraft, the *Terror*, that could travel at speeds of over 200 miles per hour and against which there was no defense. In 1907 Wells wrote *The War in the Air*, which depicted horrific aerial bombing of cities that led to the total collapse of civilization. Here is an excerpt discussing the strategic bombing of New York City: "She was the first of the great cities of the Scientific Age to suffer by the enormous powers and grotesque limitations of aerial warfare. She was wrecked as in the previous century endless barbaric cities had been bombarded, because she was at once too strong to be occupied and too undisciplined and proud to surrender in order to escape destruction. Given the circumstances, the thing had to be done. It was impossible for the Prince to desist, and own himself defeated, and it was impossible to subdue the city except by largely destroying it. The catastrophe was the logical outcome of the situation, created by the application of science to warfare. It was unavoidable that great cities should be destroyed."[33] In 1914 Wells published *The World Set Free*, which predicted the combination of two new types of warfare—aerial bombing and "atomic" weapons—used in a catastrophic war.

These works and others fostered a sense of "air-mindedness" and planted the seed for the idea that airpower could be decisive and pivotal (which became problematic for the constraining strategic bombing norm later on).[34] However, this was still solidly the realm of fiction, and while it informed the ongoing debates over the strategic bombing norm, it was not a sufficient catalyst to motivate a great many norm entrepreneurs and leaders or generate domestic political pressure for action. The only widely reported demonstration of airpower prior to World War I was the Italo-Turkish War of 1911–12, when an aerial bomb hit a Turkish hospital, probably inadvertently. This elicited howls of protest, such as Bertha von Suttner's assertion that it represented the "barbarization of the air."[35] This incident and the widespread fiction depicting future strategic bombing attacks led two hundred prominent British citizens to sign a statement calling for the prohibition of aerial weapons; unfortunately, the Second Hague Conference was the last major effort and opportunity to codify the emerging strategic-bombing norm.[36] The third Hague Conference was canceled when World War I broke out.

Lack of Clarity and Inadvertent Violations

While norm entrepreneurs had achieved some progress in spreading the nascent strategic bombing norm prior to World War I, once war began the lack of clarity (a factor identified by norm evolution theory as important for norm development) and limits of the technology that led to inadvertent violations weakened the norm's constraining force. World War I, begun in July 1914, was the first real test of the emerging strategic bombing norm as it was the first major conflict that occurred following the advent of tangible airpower capable of delivering some form of munitions from the air. Given the limited number of norm entrepreneurs and leaders and confusion regarding aspects of the norm (such as what constitutes a defended city or a legitimate military target), it is no surprise that the norm was not effective in fully constraining aerial bombing during the conflict. Initially, President Woodrow Wilson and Pope Benedict XVI reached out to the belligerents to encourage them to avoid bombing cities.[37] This request by these norm leaders was largely heeded, and during the war most use of airpower was oriented toward battlefield support to ground forces, known as tactical airpower.[38] For example, in January 1915 the kaiser authorized attacks only on costal facilities and military targets on the coast of Great Britain in order to comply with the codified strategic bombing norm.[39] However, strategic bombing of nonmilitary targets did occur. In June 1915 the Germans were the first to breach the emerging norm by conducting zeppelin attacks against London and the nearby coastal areas, although these attacks achieved little effect.[40] The Germans deliberately violated the norm only as a response to inadvertent British attacks on German cities. The French and others responded against industrial and military targets behind the lines of control. Given the lack of precision and the frequent need to bomb at night, these aerial bombing attacks regularly, albeit inadvertently, struck civilian targets, leading each side to perceive the attacks of the other as indiscriminate (while their own were viewed as legitimate).[41] This misperception due to limited experience and understanding of the technology helped undermine the emerging norm constraining strategic bombing. Norm evolution theory indicates norms may emerge more rapidly for a new issue area or technology as prevailing norms are not yet well established; however, this experience also shows that a new area may be rife with misperceptions that can hamper norm emergence.

In 1917 the Germans escalated their strategic bombing offensive and incorporated a new weapon into their arsenal: the Gotha series bombers. The Gotha bomber aircraft were massive twin-engine biplanes with a wingspan slightly shorter than a World War II B-29 and a capacity to carry up to 1,500 pounds of bombs.[42] Air raids against London using the new Gotha bombers were more effective and led to heightened public outcry.[43] The London *Times* published a "reprisal map" of German towns within 150 miles of the front lines, and public meetings were held where resolutions were passed "amid great cheering" calling on the British government to retaliate against German cities.[44] In response to this domestic pressure, the British established the Royal Air Force (RAF) and sanctioned reprisal strategic bombing attacks against the Germans. The RAF essentially became a potent new norm entrepreneur for strategic bombing, advocating for permissive rather than constraining norms. Maj. Gen. Hugh Trenchard became the head of the new RAF; although he was unable to achieve much success against the Germans due to limited resources and technology, he began to frame strategic bombing in a way that would prove toxic to the constraining norm in the interwar years. He emphasized the "morale effect" of strategic bombing and offered the soon famous statistic that "the morale effect of bombing stands undoubtedly to the material effect in a proportion of 20 to 1."[45] The basis of the statistic was dubious and civilian morale particularly difficult to measure or quantify, but the notion stuck.[46] Interestingly, near the end of the war, perhaps borne out of a material interest to stop adversary strategic bombings, the German government expressed interest in a mutual agreement with the Allies to cease strategic bombing and in October 1918 attempted to move back to its 1914 policy of bombing only the battlefield.[47] However, the Allies did not revert to full adherence to the constraining norm, although on May 30, 1918, they refrained from any bombing attacks so the Germans could celebrate a religious holiday.[48] While the emerging constraining norm on strategic bombing prevented extensive deliberate aerial targeting of civilians and nonmilitary targets, limits in technology and military organizational interest also deserve credit for the strategic bombing restraint. The true test of the norm was yet to come.

Unproven Technology and Competing Humanitarian Perspectives

Norms never emerge in a vacuum and are always competing with other norms to define acceptable behavior. While the strategic bombing norm had achieved its early success due to coherence and grafting onto the norm of noncombatant immunity, that would impede its progress during the interwar years. As strategic bombing was an unproven technology, a competing and permissive norm emerged based on the hypothesis (grounded in its coherence with the norm of noncombatant immunity) that strategic bombing would end wars quickly or prevent them altogether and thus was more humanitarian. As a case in point, Wells's *The World Set Free* in 1907 predicted that "with the flying machine war alters its character; it ceases to be an affair of 'fronts' and becomes an affair of 'areas'; neither side, victor or loser, remains immune from the gravest injuries."[49] While the experience of World War I was somewhat underwhelming in terms of airpower's effectiveness, ideas such as Wells's took hold during the interwar years. For example, despite the failure of the Zeppelin attacks to cause much damage in Britain, people remembered feeling vulnerable and defenseless.[50] The idea that strategic bombing could prove decisive by effectively degrading enemy morale—Trenchard's 20-to-1 assertion—and capacity for fighting only grew, although, to be fair, Trenchard did soften his statements about airpower's efficacy by saying that he did not mean to "imply that the air by itself can finish the war."[51] Liddell Hart's *Paris, or the Future of War*, published in 1925, was widely read and further instilled the notion of airpower's effectiveness with scenes of "women, children, babies in arms, spending night after night huddled in sodden fields" and his assertion that a strategic bombing attack on half a dozen great cities would result in the evaporation of the general will to fight.[52] The morale argument in particular was compelling at the time given unrest in working-class urban environments, such as the British general strike in 1926 and Paris riots in 1934.[53] This perception of a fragile civilian target fed the airpower advocates' argument that strategic bombing would quickly lead to major unrest and devastation.

In addition to popular fiction like Wells's, military airpower theorists proved to be some of the strongest advocates for an increased role for aircraft in armed conflict. The British author J. M. Spaight was one of the first proponents of airpower, arguing that "no amount of composure, no surplusage of bull-dog tenacity can save a people raided copiously, scientifically, systematically."[54] Brig. Gen. Guilio Douhet's *Command of the Air* was published in 1921 and argued strongly that strategic bombing campaigns would prove decisive and devastating in future conflicts. In this sense Douhet and other like-minded airpower theorists were actively working to undermine the constraining norm against strategic bombing. However, their arguments were actually connected to the established norm on which the strategic bombing norm was trying to graft: noncombatant immunity. Douhet and others argued that due to the impact of strategic bombing on national morale and its ability to bring the conflict to cities and civilians, it would actually make war shorter with less overall noncombatant harm and damage. To this point he said "Mercifully, the decision will be quick in this kind of war, since the decisive blows will be directed at civilians, that element of the countries at war least able to sustain them. These future wars may yet prove to be more humane than wars in the past in spite of all, because they may in the long run shed less blood."[55] This was not an entirely new idea—Maj. J. D. Fullerton of Great Britain had argued as early as 1893 that the presence of enemy airships over a capital city would end the conflict—but it took on newfound interest and prominence.[56] This increased prominence was in part due to the fact that air forces were attempting to play up their potential role in future conflicts as part of the organizational competition for scarce resources within military organizations. Others theorists, such as Spaight and the American Billy Mitchell, echoed Douhet's views (although Spaight later recanted them and promoted the constraining strategic bombing norm).[57] Mitchell in particular, advocating unsuccessfully for an independent air force in the United States, made the claim—based on this theoretically decisive role—that airpower and strategic bombing had revolutionized war and made armies redundant.[58]

By the 1930s technology began to show signs of promise toward achieving the deadly capabilities underlying Douhet's vision. Aircraft shifted from wooden biplanes to metal monoplanes with increased distance and hauling capacity.[59] These advances as well as limited experience with anti-aircraft capabilities led many to believe that defense was not an effective strategy against modern strategic bombing and that bombers would always reach their target.[60] It is important to note that the post–World War I context also bolstered the "humanitarian" argument for strategic bombing advanced by Douhet and others. Given the widespread use of chemical weapons, many assumed they would be part of the deadly cocktail of bombs dropped on cities, leading to horrifying and devastating results. Coupled with the perceived fragile social structure in urban society and the view that "the bomber will always get through," it is understandable that many accepted the claim that strategic bombing would actually shorten the war.

Organizational Platforms Emerge and Spread

Following World War I, efforts were launched to utilize new international organizations and events to codify various norms for warfare and disarmament so as to prevent another great war. These served as organizational platforms for norm entrepreneurs to support the strategic bombing norm. Norm evolution theory identifies these platforms as a key element that makes successful norm emergence more likely. The Washington Naval Conference of 1921–22 was one such venue; it was successful in reaching an agreement prohibiting first use of chemical weapons.[61] The delegates also entertained a proposal to codify the constraining norm on strategic bombing by limiting the number or characteristics of military aircraft. They ultimately gave up due to the fact that commercial aircraft could not readily be limited or restricted and could easily be converted to military purpose—a "dual-use" problem.[62] Instead, delegates decided to hold a separate conference dedicated to regulating airpower during war. This conference was held in December 1922 in The Hague. The delegates grappled with appropriate targets for aerial bombing and ultimately agreed to rules reasserting the constraining strategic bombing norm and the concept of noncombatant immunity.[63] In reference to the aerial bombing conducted during World War I, they stated, "The conscience of mankind revolts against this form of making war outside the actual theatre of military operations, and the feeling is universal that limitations ought to be imposed."[64] The key aspects of the proposed rules were Articles 22 and 24: "Aerial bombardment for the purposes of terrorizing the civilian population, of destroying or damaging private property not of a military character, or of injuring non-combatants is prohibited" and "Aerial bombardment is legitimate only when directed at a military objective, that is to say, an object of which the destruction or injury would constitute a distinct military advantage to the belligerent."[65] Unfortunately, due to uncertainty that still remained regarding aircraft technology and the grand claims of airpower theorists, no nation adopted the proposed rules.[66] But even had they been adopted, they would not have clearly settled the matter due to the ambiguity of the term *military objective*. John Bassett Moore, a U.S. participant at the Hague conference, emphasized this point: "The doctrine of the 'military objective' . . . will be no adequate protection. Any belligerent who chooses will be able to keep within the rules which embody that doctrine and yet use his air arms for a purpose quite distinct from the destruction of objects of military importance, namely, for the creation of a moral, political, or psychological effect within the enemy country."[67]

Multiuse Technologies and State Self-Interest and Mistrust

In addition to the effort to come up with an international agreement to constrain strategic bombing by codifying the emerging norm, there were even more ambitious proposals to completely remove airpower from national control. However, the multiuse nature of bombing technology along with state self-interest and mistrust impeded norm development. In the 1920s the French government offered the Bourgeois Plan to establish a multinational peacekeeping force, including a multinational air force, under the League of Nations.[68] Similar to President Eisenhower's initial approach to nuclear weapons as part of his Baruch Plan, the idea was supported by figures such as Spaight and Winston Churchill.[69] Ten years after the Hague conference, this idea became intertwined with one more attempt to solidly and codify the constraining norm for strategic bombing with the Geneva Disarmament Conference of 1932. Airpower theorists and aircraft technology had limited the emerging norm's prominence and was contesting its spread. As was the case with chemical weapons, these lobbying efforts for resources for air forces (either as their own military service, as was the case with Britain, or as a component of another service, the case in the United States) had an unintended secondary effect of stoking public fear of strategic bombing. This growing fear and concern led to diplomatic talks in Geneva in 1932, where delegates again were trying to come up with a way to constrain military air forces in the context of the dual-use threat posed by commercial aircraft. This was perhaps the best chance for reaching a genuine consensus.[70]

Early into the conference the French proposed the transfer of all military aircraft to the League of Nations and a total prohibition on national airpower.[71] While the smaller nations supported the proposal, the great powers did not. The discussion then moved to a general prohibition on strategic bombing akin to what was temporarily adopted at The Hague in 1899. Perhaps foolishly the British would agree to the prohibition only if it allowed strategic bombing "for police purposes in certain outlying areas," namely its colonies.[72] The British insistence on this as based on their belief—fueled by individuals such as Trenchard—that strategic bombing was invaluable in maintaining peace in their colonies. These differences became irrelevant in October 1933 when the German delegation walked out of the conference and simultaneously withdrew from the League of Nations, ending the best hope that further disarmament efforts or wartime regulations would be enacted.[73] This development was not indicative of direct German opposition to the constraining norm on strategic bombing but rather was a general response against disarmament efforts as Hitler began to build up the German war machine. Just a few years later, in 1935, Hitler expressed an interest in a "prohibition on indiscriminate bombing of densely populated regions," and in 1936 the German government unsuccessfully began planning for another conference to revisit the issue.[74] The British were extremely interested (and even contemplated offering accession to German remilitarization of the Rhineland as an inducement) but ultimately could not pursue the deal without French participation, which was lacking.[75] Ultimately, while there remained a broad and general interest in negotiating some constraints on strategic bombing, the mistrust and divergent perspectives of the major powers prevented such an effort from ever succeeding. Similar to the period before the Hague Conference of 1907, each nation's air force was at a different state of readiness and their focus on strategic bombing varied. The British feared it greatly as it made their island nation suddenly vulnerable, while the United States was less interested in major constraints as, in the eyes of President Franklin Roosevelt, a bombing campaign with relatively few casualties might be acceptable to American isolationists.[76]

The Interwar Period

During this interwar period there were some smaller-scale conflicts with indigenous uprisings in various colonies and the civil war in Spain, which served as a military testing ground and proxy war for the major powers. Many lessons regarding airpower and strategic bombing arose from these experiences, and the demonstrations fueled altruism and empathy motives in support of the strategic bombing norm. In colonial territories such as India and Iraq, the major powers saw that airpower alone could subdue rebellions through "terror bombing tactics."[77] For example, in 1922 the British turned over control of Iraq to the RAF, which used indiscriminate aerial bombing to police the territory with a minimal number of ground troops.[78] The British and the United States were not directly involved in the Spanish Civil War and largely discounted observations that did not match their existing perceptions and notions of airpower.[79] The other major powers were overtly or covertly involved in supporting one of the two sides. Most of the aerial attacks in Spain were tactical in nature and oriented against military forces and targets.[80] However, cities were not immune from the conflict. The most noteworthy and recognized demonstration of strategic bombing occurred in Guernica on April 26, 1937.[81] The aerial attack by the German Condor Legion on the Basque town is considered one of the first large-scale strategic bombing raids and led to many casualties and hundreds of deaths. It also was the inspiration behind Pablo Picasso's famous painting *Guernica*, which vividly depicts the violence of the bombing. The painting shows a woman grieving over her dead child, a horse falling in agony, trampled and distraught bodies, and a man raising his arms in terror, his right hand depicting the shape of an airplane. The Guernica attack, perhaps more than any other, stoked fear of strategic bombing and fueled public perception that airpower could rapidly degrade morale and possibly end conflicts sooner. The *New York Times* reporter Herbert Matthews, who witnessed the strategic bombing in Spain, wrote, "Human beings are not built to withstand such horror. . . . [It] makes one either hysterical or on the verge of hysteria . . . hard to remain sane."[82] This was a demonstration that undermined rather than bolstered the constraining strategic bombing norm as it fueled Douhet's argument that strategic bombing in a major war between great powers would be so terrible as to quickly bring the war to an end. This experience also helps illustrate how demonstrations of emerging-technology weapons do not necessarily always bolster a constraining norm and can sometimes weaken it. While numerous variables dictate which impact (positive or negative) a demonstration will have on a constraining norm, the demonstration does create a window for movement (either further development or erosion) for the norm.

Socialization and institutionalization are key actor mechanisms norm evolution theory identifies as important to norm emergence and eventually achieving a norm cascade. As a consequence of the failures of the 1920s and 1930s to achieve a tipping point and codify the strategic bombing norm, just prior to World War II the norm was not well internalized or socialized by the armed forces of the major nations. At that time Britain's *Manual of Military Law* stated, "No legal duty exists for the attacking force to limit its bombardment to the fortifications or defended border only. On the contrary, destruction of private and public buildings by bombardment has always been, and still is, considered lawful, as it is one of the means to impress upon the local authorities the advisability of surrender."[83] The German Luftwaffe's document *The Conduct of the Air War* said it was the armed forces' duty to take the war to the "enemy's homeland [and to] attack his military power and the morale of the enemy population."[84] The air force personnel of the major powers were largely aware of the rules developed, although not adopted, at the 1923 conference in The Hague, but they clarified that they were not legally binding.[85] National leadership did seem to declare support for the norm on the eve of conflict, with Hitler announcing that he would restrict Luftwaffe bombing to military targets and Prime Minister Neville Chamberlain declaring that Britain would "never resort to the deliberate attack on women and children, and other civilians for the purpose of mere terrorism."[86] While these statements made them appear to be norm leaders and appealed to the altruistic and humanitarian perspectives of those supporting the norm, both allowed—perhaps inadvertently—ample wiggle room for future violations of the norm. Hitler did not specify how a "military target" would be defined, and Chamberlain's declaration that Britain wouldn't bomb noncombatants included the caveat "for the purpose of mere terrorism." Implied was that they could be bombed for other reasons. These statements are representative of the norm's lack of clarity and specificity, factors that norm evolution theory indicates are important for a norm to emerge and reach a tipping point. The ambiguity in the strategic bombing norm as constructed—coupled with technical limitations—made it particularly brittle.[87]

The Norm Collapses during World War II

Once war began in 1939, President Roosevelt, acting as a powerful norm entrepreneur, appealed to the Germans, British, and French to avoid strategic bombing and limit aerial bombardment to military targets.[88] All three agreed but stated that they would respond in kind if the other parties ignored the agreement. For the first nine months of the war they avoided bombing each other's cities and innocent civilians.[89] This was a "convention-dependent" norm, sustained only as long as mutually observed.[90] Interestingly the Luftwaffe arguably indiscriminately bombed Warsaw and Rotterdam in early 1940 and faced no retaliatory strategic bombing from the RAF or French Air Force (the French would surrender soon thereafter).[91] This seemed to indicate an unwillingness to violate the norm if the victims of the belligerent action were a third party. On May 15, 1940, the British war cabinet made a fateful decision to conduct aerial attacks of oil and railway targets in the Ruhr, beyond the Rhine.[92] The British anticipated an immediate German response on British towns, but none came. The lack of a German response has been attributed to the organizational culture of the Luftwaffe, which was tactically oriented and did not place much faith in strategic bombing.[93] The RAF, however, had a different organizational culture. When the aerial Battle of Britain began and Luftwaffe bombers accidentally bombed London, the British immediately responded and RAF Bomber Command (which was predisposed against the constraining norm) ordered a strategic bombing attack on Berlin the following night.[94] However, even with this tit-for-tat a full-scale embrace of strategic bombing did not immediately occur, and both sides continued to pursue primarily military targets.

Soon, though, it became apparent that aerial bombing had limited accuracy, especially when bombing at night to avoid catastrophic losses in the bombing force. A British photo-reconnaissance study, known as the Butt Report, reported that "of those aircraft recorded as attacking their target, only one in three got within five miles."[95] Upon seeing this report Churchill famously declared, "The most we can say is that [bombing] will be heavy and I trust seriously increasing annoyance [for the Germans]."[96] The Butt Report conflicted with the anecdotal claims of accuracy, such as when Sir Richard Peirse, head of British Bomber Command, said, "I don't think at this rate we could have hoped to produce the damage which is known to have been achieved."[97] Faced with more empirical data, such as the Butt Report, Peirse's anecdotal perspective gave way to a more realistic understanding of the accuracy of strategic bombing. Large cities and urban centers were much easier targets to hit, and thus, as the war continued, strategic bombing became more palpable as public pressure to respond in kind grew and came to be viewed as the most effective employment of bombers. As a result, in 1942 the British Bomber Command issued a directive stating that operations would shift to focus on "the morale of the enemy civilian population and in particular, of the industrial workers."[98] Technically this could be interpreted as compliant with the norm codified in the unadopted declaration from the Hague Conference of 1922, which identified legitimate military targets as those whose destruction or injury would constitute a distinct military advantage to the belligerent, but such logic required such a stretch as to render almost everything a military target. Soon British Bomber Command decided to pursue a policy of "de-housing" by explicitly striking German residential targets.[99] The strategic bombing targets identified by airpower theorists such as Douhet were now the prime focus of bombing efforts during the war, although they did not end the war as rapidly and thus "humanely" as hoped. Even when better technology became available to allow for more restrained and discriminate bombing, the British did not change course.[100]

As World War II raged on, the normative constraint on strategic bombing completely collapsed and extensive strategic bombing ensued.[101] When the United States entered the war in 1942, they quickly agreed to a combined strategic bombing campaign with the British flying at night and the Americans during the day.[102] The United States was presumably attacking "military-industrial" targets with their acclaimed Norden bombsight, but in 1943 the decision was made to bomb using radar during bad weather, accepting a significant amount of inaccuracy and inevitable collateral civilian damage.[103] Later in the war, when technology enabled increased accuracy, the Allies did not shift back from strategic bombing to more precision targeting of military targets because they believed they were close to breaking German morale. This implosion of the regulative norm against strategic bombing of cities and civilians was overwhelmingly evident by the end of the war. On February 13–14, 1945, the Allies bombed Dresden, which at the time was crowded with refugees. The atmospheric conditions created a firestorm so intense that civilians in shelters were "cooked until their bodies were charred husks and their body fats formed a thick layer on the floor."[104] The Pacific theater also witnessed intense strategic bombing by the end of the war. For example, on March 9–10, 1945, the United States firebombed Tokyo with devastating results: "On a clear night . . . more than 300 U.S. B-29 bombers launched one of the most devastating air raids in history. By dawn, more than 100,000 people were dead, a million were homeless, and 40 square kilometers of Tokyo were burned to the ground."[105]

The end of the war put an exclamation mark on the strategic bombing norm's collapse when the United States dropped nuclear weapons on the Japanese cities of Hiroshima and Nagasaki (officially considered military targets).[106] Initial norms regarding the use of nuclear weapons were essentially a continuation of the current state of norms regarding strategic bombing, principally the absence of any constraint on strategic bombing. This was highlighted in Gen. Curtis LeMay's statement of moral comparison: "We scorched and boiled and baked to death more people in Tokyo on that night of March 9–10 [1945] than went up in vapor at Hiroshima and Nagasaki combined."[107] By the end of the war the Germans had dropped 74,172 tons of bombs on Britain, and the British and United States had dropped 1,996,036 tons on Germany and German forces elsewhere in Europe.[108]

The war tribunal at Nuremberg did not address the issue of strategic bombing. The chief counsel for war crimes noted that the constraining norm for strategic bombing "had been observed only partially during the First World War and almost completely disregarded during the Second World War. . . . Aerial bombardment of cities and factories has become a recognized part of modern warfare as carried on by all nations."[109] Ironically, following the war the U.S. Strategic Bombing Survey (USSBS) and British Bombing Survey Unit (BBSU) both concluded that heavy strategic bombing was not particularly effective in diminishing productivity or breaking morale.[110] More narrowly defined military targets were more effective at achieving victory—but that recognition came too late to recalibrate the claims of the airpower theorists and advocates. While the strategic bombing norm had begun to approach a tipping point in the early twentieth century and made inroads even during the tumultuous interwar period, World War II represented its total collapse. Soon after the war a speaker at a meeting of the American Society of International Law summarized the point when he stated, "As you are all probably aware, there are no rules governing aerial warfare."[111] Norm emergence would have to continue before a true tipping point could be achieved.

Norm Reemergence after World War II

Following the horrific excesses of strategic bombing in World War II, the constraining norm began to reemerge, although it would not change in time to constrain early nuclear bombing strategy or bombing in the Korean War.[112] This was due to the recognized failure of the competing permissive norm, which was based on the view that strategic bombing was humanitarian as it would end war quickly, as well as the increasing use of shame as an actor mechanism to promote the constraining strategic bombing norm. Decisive negative impacts on enemy morale—a central objective of strategic bombing—was a nearly universal disappointment during World War II, and urban populations proved far more resilient and adaptable than predicted.[113]

The results of the USSBS and BBSU studies also confirmed that the economic impact was less than anticipated and provided greater certainty of what was technically possible. Additionally, the development of nuclear weapons delivered by bombers, foreseen by Wells in *The World Set Free*, presented the horrific and ultimate realization of complete and total strategic bombing. While their use in World War II was generally supported by the public, their introduction into the public consciousness likely indirectly raised sensitivity toward conventional strategic bombing. After all, how could one find the prospect of a nuclear holocaust in a city unacceptable yet accept the conventional strategic bombing conducted in Dresden or Tokyo?

The first test of the reemerging norm was the Korean War in 1950. Strategic bombing with incendiary explosives killed hundreds of thousands of Korean civilians south of the Yalu River.[114] However, despite the continued use of strategic bombing by the United States, there was an increasing sensitivity to public perception and the constraining norm. Civilian U.S. leaders correctly perceived bombing noncombatant civilians as politically costly due to the ethical condemnation it would generate.[115] The Chinese, motivated by their own interests but recognizing and exploiting the rising norm, decried the U.S. strategic bombing effort as a "war against Korean women and children and the old people in the rear."[116]

Strategic Bombing Norm Cascade

In light of the strategic bombing norm's early emergence, followed by stumbles and then a collapse, the likelihood of ever reaching a norm cascade was uncertain. However, following World War II and the Korean War a series of factors identified by norm evolution theory helped the strategic bombing norm solidify and reach a tipping point. These include the optimal timing and focusing event of the global awareness of the horrors of unrestricted strategic bombing coupled with real-time television reporting of ongoing military operations. The optimal timing of bombing technology had matured to the point where precision bombing was increasingly possible, making adherence to the norm possible while still exercising military airpower. These factors facilitated the institutionalization of the norm and motivated states to seeking international legitimacy of their efforts to adhere to the norm.

Various Factors Facilitated Institutionalization

The resurgence of the constraining strategic norm was even more apparent during the Vietnam War, and it was during this time that it reached a tipping point and a norm cascade began to occur. This was largely due to optimal timing, efforts to seek international legitimacy, and improvements in technology—all of which facilitated institutionalization, a key actor mechanism identified by norm evolution theory. Early in the war the U.S. military wanted to decisively attack a long list of targets in North Vietnam in order to achieve kinetic effects and shock the morale of the Viet Cong.[117] However, recognizing the resurgence of the constraining norm and openly flouting it by engaging in such a robust strategic bombing campaign carried risks of public condemnation. So instead Secretary of Defense Robert McNamara and President Johnson pursued more selective and judicious targeting.[118] White House guidance developed in 1966 specified that pilots needed to receive detailed briefings on the need to avoid civilian casualties.[119] This sensitivity to the norm was in part an effort to seek international legitimation, as predicted by norm evolution theory.

During the war a variety of major bombing campaigns were launched, including Rolling Thunder, Linebacker, and Linebacker II.

The Rolling Thunder bombing campaign, conducted from 1965 to 1968, was referred to as "one of the most constrained military campaigns in history," and the subsequent Linebacker campaigns were equally constrained, although bombing occurred at a faster pace.[120] During Linebacker the military required the use new of precision-guided munitions (PGMs) when conducting attacks in highly populated areas. Even as Johnson and then Nixon allowed gradually more latitude in targeting, political sensitivity toward adherence to the norm remained (much to the dismay of some in the military who believed these limits blocked success). Ultimately 6,162,000 tons of bombs (far more than those dropped by the Allies in World War II) were dropped on Vietnam, with little effect in breaking Viet Cong morale.[121] Despite the increased focus on avoiding civilian casualties and adhering to the strategic bombing norm, the media criticism of the bombing campaign was intense. The *Washington Post* called the Linebacker II campaign the "most savage and senseless act of war ever visited . . . by one sovereign people on another."[122] The media, with its increasingly global and instant reach through television and critical reporting of civilian casualties, motivated more norm entrepreneurs and leaders through emotional evocation of altruism and empathy.[123] The emerging media age represented a new environmental factor of optimal timing, which norm evolution theory indicates is important to achieving a norm cascade. The Vietnam experience with the media and leadership sensitivity to domestic and international pressure—along with the advent of PGMs and other new bombing technology—spurred the U.S. Air Force to move away from overtly seeking to achieve the morale effects promised early in the airpower era.[124] The airpower expert Ward Thomas notes that from the Korean War to Vietnam, there clearly was a greater emphasis on avoiding civilian casualties, reflecting the norm's continued reemergence.[125] The

U.S. experience in Vietnam also helped lead to what became known as a revolution in military affairs (RMA) and increased emphasis on precision attacks rather than traditional strategic bombing objectives. The example of growing U.S. restraint and strong public response and outcry over civilian casualties made a strong statement to the other nations watching the conflict.[126] This further persuaded nations to adopt the norm through the actor mechanism of demonstration.

Strategic Bombing Norm Internalization

Although, like the norm cascade, it may never be reached, the third and final stage in the norm life cycle is norm internalization. This phase is the effort to codify the norm and achieve universal or near universal adherence.[127] For the strategic bombing norm, norm internalization began in the early 1970s as civilian casualties in general and specific targeting via strategic bombing became increasingly unacceptable. The norm's continued coherence with the norm of noncombatant immunity along with the persistent presence of an ever watchful television media, the involvement of law and lawyers and international bureaucracies (as diplomats and the international arms control apparatus), continued improvements in precision bombing, and conformity motivations facilitated norm internalization, consistent with the predictions of norm evolution theory.

International Bureaucracies

The constraining strategic bombing norm prohibiting the deliberate targeting of nonmilitary targets for economic or morale effects began to be institutionalized through international agreements and the UN bureaucracy. Following Vietnam, some international efforts to codify the reemerged strategic bombing norm began, including a conference in 1977 that developed what became known as Protocol I to supplement the Geneva Conventions of 1949, which, among other things, sought to further codify and institutionalize the strategic bombing norm. Protocol I was adopted in June 1977; Article 51 specifically prohibited "indiscriminate attacks" that are not directed at specific military objectives.[128] Article 51 went further to define indiscriminate attacks as

> (a) an attack by bombardment by any methods or means which treats as a single military objective a number of clearly separated and distinct military objectives located in a city, town, village or other area containing a similar concentration of civilians or civilian objects; and
> (b) an attack which may be expected to cause incidental loss of civilian life, injury to civilians, damage to civilian objects, or a combination thereof, which would be excessive in relation to the concrete and direct military advantage anticipated.[129]

These restrictions addressed the ambiguity that was present in strategic bombing agreements prior to World War II and further codified the strategic bombing norm through the actor mechanisms of diplomacy and institutionalization. As of October 2013, 163 nations have ratified Protocol I.[130] While the document is somewhat controversial, elements of it are recognized by most parties (including the United States and Russia), and it has been supported by multiple UN General Assembly resolutions.[131]

The Conformity Motive

The internalization of the strategic bombing norm was not fully uncontested, as strategic bombing did occur in ongoing conflicts between regional powers. But even when contested, the conformity motive helped maintain internalization. For example, during the

Iran-Iraq War a strategic bombing campaign that became known as the War of the Cities was launched. In February 1984 Saddam Hussein's air force struck eleven Iranian cities in order to lower Iranian morale and force Iran to the negotiating table.[132] As was the case with strategic bombing in World War II, this strategic bombing attack did not achieve its desired decisive effect but did lead to reciprocal Iranian strategic bombing and missile attacks against Iraqi cities. The strategic bombing campaign evoked international condemnation, unlike the relatively tepid response to Iraq's use of chemical weapons.[133] The UN Security Council Resolution 598, which led to the August 1988 ceasefire, included a condemnation of the strategic bombing.[134] Specifically the resolution denounced the "bombing of purely civilian population centers, attacks on neutral shipping or civilian aircraft, the violation of international humanitarian law and other laws of armed conflict."[135] This condemnation showed that the internalization of the norm through actors such as the international legal bureaucracy had come a long way since the collapse in World War II, when the prevailing view was that "there are no rules governing aerial warfare."[136] Norm internalization clearly benefited from the growing international humanitarian and arms control bureaucracy, as predicted by norm evolution theory.

Improvements in Technology and Compliance Visibility

Later conflicts continued to demonstrate the norm's internalization, which was strengthened by improvements in precision targeting technology and the awareness of compliance facilitated by the ever-present media. The Persian Gulf War in 1991 began with an aerial bombing campaign that ultimately involved over 100,000 sorties and the dropping of 88,500 tons of bombs.[137] However, due to improved technology permitting prevision targeting, such as satellite (Global

Positioning System) and laser guidance systems coupled with PGMs, the bombing campaign focused on military targets and did not result in large-scale civilian casualties.[138] This was a continuation of the trend toward increasing acceptance and diffusion of the strategic bombing norm.[139] Additionally, given the emergence of a real-time global media with television networks such as CNN, the public awareness of daily conduct—including the impact of aerial bombing—helped further disseminate and integrate the strategic bombing norm through increased public pressure for adherence in order to achieve legitimacy, reputation, and esteem (as predicted by norm evolution theory). U.S. Central Command, the unified combatant command leading the Coalition forces in the conflict, placed sensitive civilian targets on a "joint no-fire target list" and reviewed all targets for the presence of sensitive civilian sites within a six-mile radius.[140]

The North Atlantic Treaty Organization (NATO) bombing campaign in Kosovo in 1999 showed even greater internalization of the strategic bombing norm, with a tremendous effort made to avoid civilian casualties. Thomas points out that numerous lawyers reviewed each proposed target, and potentially sensitive targets, especially those in urban areas such as downtown Belgrade, were reviewed by officials in each of the nineteen NATO capitals.[141] A few years later, in 2001, the U.S. military campaign in Afghanistan, later conducted by the multinational International Security Assistance Force, demonstrated an even greater and heightened focus on preventing civilian casualties.[142] However, internalization is not complete and appears to be particularly weak during intrastate conflict. For example, during the Syrian Civil War that started in 2011, the Baathist authoritarian government publicly claimed to have generally pursued discriminate targeting and avoided bombing civilians (even while it was willing to deploy forbidden chemical weapons).[143] However, human rights groups reported that Syria conducted extensive, in some cases deliberate, aerial attacks against civilians, leading to over 4,300 deaths between July 2012 and April 2013.[144] The norm and the desire for conformity it creates are present nonetheless as Syria vocally denied these claims as part of its propaganda strategy.

This more recent demonstration of the limits of the constraining strategic bombing norm highlights that norm internalization can vary among the different elements of the international community (for example, Western democracies vs. authoritarian regimes) and different types of conflict (such as interstate war vs. intrastate war) as it moves toward or away from near universal internalization. Syria aside, the demonstrated adherence to the strategic bombing norm during some of these more recent conflicts indicates that it has achieved partial internalization and key actors such as lawyers and diplomats, airmen, and the international bureaucracy have fully embraced the norm. It is now more clear and specific than during the first half of the twentieth century, and the optimal timing of PGM military technology and the presence of a global real-time media have further deepened its adoption, as predicted by norm evolution theory. Eliot Cohen summarized the current state of the norm, pointing out that "it is highly unlikely that advanced powers will again resort to the wholesale devastation of cities and towns, whether to shatter enemy morale or as a by-product of efforts to hit other target systems, such as railroad yards or factories."[145]

Summary

As with chemical and biological weapons, the emergence, cascade, and internalization of the constraining norm for strategic bombing were in most aspects consistent with the various hypotheses from norm evolution theory. The emergence of norms for strategic bombing arose through coherence (connections to key ideas, adjacency, and grafting) with the long-established norm of noncombatant immunity. This connection proved problematic as technology and the emerging industrialized "total war" environment would make application of this concept extremely difficult. The norm arose before true airpower had even been invented, leading to imaginative speculation as to its capabilities and impact. Spaight alluded to this phenomenon when he said that the bomber's "mystery is half its power."[146] While the preemptive emergence of the norm offered an opportunity to lock in a prohibition before the technology matured, as was done with chemical weapons, the countervailing speculation that strategic bombing could actually make wars more humane by shortening their duration severely weakened the preemptive proscription by limiting it to just a few years.[147] Limited demonstration of strategic bombing during World War I, coupled with global insecurity and fears of future aerial chemical weapon attacks, were insufficient to serve as a catalyst for a norm cascade following the war. This allowed more time for continued speculation and capability inflation by airpower theorists, science fiction authors, and the press, as well as the creation of airpower bureaucracies that in many cases bolstered support for permissive norms. Further, the state distribution of power, level and diffusion of aerial technology, and natural resources and boundaries (such as Britain's previously highly defensive position as an island nation) created divergent interests in cultivating and codifying the emerging norm during this interwar period. As a result the constraining norm on strategic bombing held out only briefly during the early days of World War II and by the end of the war had collapsed. It took reflection on the horrific humanitarian impact of strategic bombing and its limited effectiveness as a war-winning tool as well as the emergence of more discriminate bombing technology and a real-time global

Table 6. Hypotheses of norm evolution theory applied to strategic bombing

HYPOTHESES	STRATEGIC BOMBING
NORM EMERGENCE	
Significant presence of norm entrepreneurs/norm leaders with organizational platforms	XX
Key actors motivated altruism, empathy, and ideational commitment	XX
Actors employ persuasion, framing, memes, shame	XX
Domestic political concerns	X
States seeking international legitimation	X
State distribution of power, prevailing levels of technology, availability of natural and human resources, etc. favoring the norm	XX
Large-scale turnover of decision makers	
Recognized failure of norms from a previous norm generation	XX
Emergence of a new issue area where prevailing norms not well established	X
Norm prominence	X
Norm coherence—connections to key ideas, adjacency, and grafting	XX
NORM CASCADE	
Key actors are involved, specifically states, international organizations, and transnational networks	X
Key actors motivated by legitimacy, reputation, esteem	XX
Actors employ persuasion and diplomacy	X
Actors employ institutionalization	XX
Actors employ socialization	X
Actors employ demonstration	X
Optimal timing	XX
Norm clear and specific	X
Norm makes universal claims	X
Norm coherence	X
NORM INTERNALIZATION	
Key actors are involved, specifically law and lawyers, professions, and bureaucracy	XX
Actors motivated by conformity or domestic turmoil	XX

Table 6. (*continued*)

HYPOTHESES	STRATEGIC BOMBING
Actors employ habit and institutionalization	X
Existence of professional networks	
Norm clear and specific	X
Norm makes universal claims	X
Norm coherence	X

More prevalent variables are indicated by XX rather than X.

media to empower the norm's reemergence and eventual cascade during the 1960s and 1970s. Today the norm is being internalized due in part to strong international pressure for conformity enabled by this real-time media. One lesson of the strategic bombing norm's experience in the twentieth century is that even a total collapse of an emerging norm does not equate to its permanent demise and that changing circumstances can help the norm reemerge and grow. Table 6 summarizes which hypotheses from norm evolution theory were present in the history of the evolution of norms for strategic bombing.

While overall the hypotheses of norm evolution theory are consistent with the experience of the constraining strategic bombing norm's life cycle, the norm development experience offers some further insights to help expand the general hypothesis of norm evolution theory for emerging-technology weapons. Specifically the experience of the strategic bombing norm highlights the challenges for norms in a new issue area where prevailing norms are not well established. Norm evolution theory indicates that this environmental factor should facilitate norm emergence by creating space for the norm to grow; however, the strategic bombing experience shows how this can also impair norm development. Actors may be unwilling or cautious about adopting norms for weapons whose effects and capabilities are uncertain. The lack of history and understanding of the limits of the new technology create opportunities for miscalculation and inadvertent norm collapse (as occurred in World War I) as well opportunities for norm entrepreneurs and bureaucracies to foster competing permissive norms (for example, the concept that strategic bombing was humane as it would end wars quickly).

4 Norm Evolution for Nuclear Weapons

> The new thermonuclear weapons are tremendously powerful; however, they are not . . . as powerful as is world opinion today in obliging the United States to follow certain lines of policy.
> President Dwight D. Eisenhower

Nuclear weapons and cyber weapons are forms of nonconventional weapons that share many characteristics and have significant international security implications, such as the potential for major collateral damage or unintended consequences (due to fallout, in the case of nuclear weapons) and the frequent use of covert programs to develop such weapons. Nuclear weapons of the early Cold War era could be considered emerging-technology weapons, as cyber weapons are today. Accordingly examining norm evolution for nuclear weapons is a valuable way to test and possibly help extend norm evolution theory and develop expectations for how norms for cyber warfare will develop.

Along with chemical and biological weapons, nuclear weapons are often categorized as WMD. National Defense University's 2006 study "Defining 'Weapons of Mass Destruction'" documents the term's emergence and over forty definitions and notes that upon their invention nuclear weapons were immediately included as WMD.[1] Nuclear weapons are explosive devices based on fission or combined fission and fusion nuclear reactions that release a tremendous amount of energy. The first nuclear weapon tested created an explosion equivalent to the detonation of approximately 20,000 tons of TNT.[2] Modern nuclear weapons are more powerful by an order of magnitude. As Thomas Schelling notes, the rapid emergence of norms against the use of nuclear weapons was so effective in constraining action that President Eisenhower's secretary of state, John Foster Dulles, when contemplating the use of nuclear weapons in 1953, said, "Somehow or other we must manage to remove the taboo from the use of [nuclear] weapons."[3] The norm against the employment of nuclear weapons was so strong that President Truman did not use them against Chinese troops during the Korean War, President Nixon did not use them in Vietnam, the Soviet Union did not use them in Afghanistan, and Israel did not use them in the 1973 war with Egypt—all circumstances where the opponent lacked such weapons and their use made some degree of utilitarian sense. Nonuse by the superpowers when facing each other in heated crises, such as the Cuban Missile Crisis, further demonstrated the norm's potency.[4] How did this norm emerge and rapidly reach a norm cascade and begin the process of internalization? The nuclear norm experience offers some particularly valuable lessons for norm evolution theory, including characterizing a weapon type as "unconventional," the potential impact of power politics on norm evolution, and the fact that norms regarding use may be easier to internalize than those regarding development, proliferation, and disarmament.

Nuclear Weapon Norm Emergence

The emergence of constraining norms for nuclear weapons was consistent with norm evolution theory. Initial permissive nuclear norms, due to their coherence (compatibility with existing norms with which they were not directly competing), arose from existing norms for strategic bombing; early on it appeared that nuclear weapons would simply be treated as bigger bombs. The involvement of key actors and norm entrepreneurs, initially the Soviets and the UN and later international groups of scientists and citizens concerned about nuclear weapons, helped utilize some of the key mechanisms identified in norm evolution theory and reframe the way people thought about these weapons and corresponding norms regarding their use.

Coherence with Strategic Bombing Norms

Although nuclear weapons were a new invention, norms regarding their possession and use quickly developed due to coherence with and grafting onto the existing normative environment of World War II. Before President Truman and his national security team even grappled with the question of employing these new weapons against the Japanese, they had already dealt extensively with the issue of strategic bombing.[5] Initial norms regarding the use of nuclear weapons were essentially a continuation of the norms for strategic bombing, which permitted gruesome firebombing of Japanese and German cities. Nowhere is this point made clearer than in Gen. Curtis LeMay's statement about scorching, boiling, and baking more people in the fire bombing of Tokyo in March 1945 than died in the two nuclear bombings of Hiroshima and Nagasaki.[6] Nina Tannenwald, a leading scholar on nuclear norms, suggests that this view was one of two major factors that led to the use of nuclear bombs against Hiroshima and Nagasaki.[7] This connection was so potent that some have referred to it as a "seamless web."[8] According to Tannenwald, the "accumulated barbarities" of World War II, such as the Nazi-perpetrated murder of 6 million Jews and Japanese attacks in Nanking, also explain why the initial use of nuclear weapons was considered logical and went largely unquestioned.[9] The historian Barton Bernstein characterizes this phenomenon as a "redefinition of morality."[10] These factors and the coherence of the initial, permissive nuclear norm with the norm for strategic bombing resulted in nuclear weapons initially being perceived as one weapon in the war-fighting tool kit.

Norm Entrepreneurs and Leaders

Norm evolution theory highlights the importance of norm entrepreneurs with organizational platforms as well as the advocacy of norm leaders (states) in helping a new norm emerge and predicts that norm emergence is more likely if more actors are involved. Initially the Soviets and the UN motivated by power-balancing politics and later international groups of scientists and others motivated by altruism, empathy, and ideational commitment (as predicted by norm evolution theory) served as key norm entrepreneurs and actors supporting a constraining norm for nuclear weapons. U.S. support for a constraining norm grew, as did support from key leaders. These key leaders played a critical role in cultivating the moral revulsion underlying the norm at just the right moments; U.S. Defense Secretary Robert McNamara played a similar key role in disseminating the idea that nuclear weapon use was unthinkable. However these developments took time and were not present in 1945. Given the permissive normative context in which nuclear weapons emerged, it is understandable that strong advocates for a constraining nuclear norm did not arise right away. Although some leaders did express concerns regarding the employment of nuclear weapons against civilians, they were in the vast minority in the early nuclear era. For example, Gen. George Marshall, the army chief of staff, advocated for the new weapons to be used only against military installations or, if used against targets where civilians were present, to provide advance notice.[11] His colleague, Admiral William Leahy, went further, framing the use of nuclear weapons in the context of heinous chemical weapons and advocated against their use.[12] Leahy's use of framing—a key mechanism identified by norm evolution theory—was unsuccessful at the time but ultimately proved to be effective. A minority of the scientists working on the Manhattan Project were also opposed to the use of the atomic weapon.[13] This lack of widespread nuclear norm entrepreneurs and leaders continued in the postwar era as the idea of using nuclear weapons received broad popular support (86 percent of those surveyed in the United States in the war's aftermath supported their use) and the antinuclear weapons movement was terribly weak.[14] However, there were signs of the seeds for the constraining norm in Leahy's arguments and in the emergence of a network of concerned scientists from the Manhattan Project who formed the Federation of American Scientists (FAS).

It was not until the late 1940s and early 1950s that a norm constraining nuclear weapons began to emerge.[15] As hypothesized by norm evolution theory, this new norm, competing to supplant the permissive nuclear norm that had grafted onto norms for strategic bombing, gained momentum as it garnered more support from norm entrepreneurs and actors working through organizational platforms. The United States, as the world's only nuclear power, shouldered the task of grappling with nuclear weapons and developing a nuclear strategy.[16] At first, recognizing the unique power and dangers posed by nuclear weapons despite its exclusive monopoly at the time, the United States pursued a goal of international control of nuclear weapons. This aspiration was initially proposed in a November 1945 declaration from the United States, Great Britain, and Canada and ultimately became known as the Baruch Plan.[17] This plan would have entailed the United

States giving up its nuclear weapons once all other nations agreed not to pursue them and an inspection regime to prevent their development was put in place.[18] This proposal was an outgrowth of a report from March 1946 developed under the leadership of Dean Acheson and David Lilienthal (and thus often referred to as the Acheson-Lilienthal Report). The Acheson-Lilienthal Report offered details of how to implement the Baruch Plan by establishing international ownership of all fissile material and the entire nuclear fuel supply chain, from uranium mines to nuclear waste.[19] Obviously such a proposal would have rapidly produced and simultaneously legally codified a norm against state possession of nuclear weapons. The ability to even contemplate such rapid norm emergence was due to some key environmental factors identified by norm evolution theory—namely the fact that the distribution of power and technology in this area favored the United States and put it into a position where it could embrace such an approach. The Soviet Union, which was aggressively pursuing its own nuclear weapons program, rejected the Baruch Plan out of a concern that the UN was stacked in support of the United States and its Western allies and refused to submit to the proposed unrestricted compliance inspections while the United States still possessed nuclear weapons.[20] However, the Soviets did support efforts to denounce nuclear weapons and statements in pursuit of nuclear disarmament. They joined with the United States in the new UN General Assembly to establish the UN Atomic Energy Commission, charged with working toward the "elimination from national armaments of atomic weapons."[21] Later, Andrei Gromyko, the Soviet representative to the new UN Atomic Energy Commission, began lobbying for "an international convention prohibiting the production and employment of weapons based on the use of atomic energy."[22]

These Soviet antinuclear efforts were largely intended to delegitimize U.S. nuclear weapons (despite the Soviets themselves acquiring such weapons in 1949) as part of the great power politics in the early Cold War era.[23] This was similar to how early advocates of the constraining norm for strategic bombing attempted to use the norm to constrain adversaries and erode their bombing advantage. This was not one of the usual motivations indicated by norm evolution theory, but Soviet efforts did help spawn such motivations in other actors in the early antinuclear weapons movement. This is a potential area where norm evolution theory could be extended to account for the fact that power-balancing considerations can inadvertently help encourage norm emergence. After the failure of the UN proposals to eliminate state-owned nuclear weapons, norm entrepreneurs with organizational platforms began to pursue a comprehensive norm against use and possession of nuclear weapons, initially through the Soviet-led World Peace Council (WPC). The WPC was an international coalition led by the International Department of the Central Committee of the Soviet Communist Party.[24] On March 15, 1950, it approved the Stockholm Peace Appeal, a petition calling for a total ban on nuclear weapons.[25] After a large public outreach effort, allegedly hundreds of millions of people around the world signed the petition, which demanded "outlawing of atomic weapons as instruments of intimidation and mass murder of peoples" and stated, "Any government which first uses atomic weapons against any other country whatsoever will be committing a crime against humanity and should be dealt with as a war criminal."[26] While the bureaucratic normative environment—born out of World War II's strategic bombing experience—was not favorable to the constraining nuclear norm, this grassroots public effort tapped into the prewar context and fear of nuclear weapons. As early as 1914 the motivating ideational commitment and altruism needed to motivate norm entrepreneurs existed, spurred by, among other things, H. G. Wells's *The World Set Free*, which predicted "atomic" weapons and described a catastrophic war in which they were used. These new norm entrepreneurs tapped into this fear by shifting the framing of the nuclear threat, using the power of nuclear weapons as seen in Hiroshima and Nagasaki.

Organizational Platforms Emerge and Spread

Norm evolution theory indicates that organizational platforms, used by norm entrepreneurs and leaders to promote their norms, are important for norm emergence and can provide different types of resources and advocacy networks to secure the support of state actors. The Cold War nuclear posturing helped turn the WPC (as a transnational network) and the UN Atomic Energy Commission (as an international organization) into organizational platforms and incubators for norm entrepreneurs and helped the budding nuclear norm emerge. Soon other networks and groups began to emerge. The international Pugwash group of scientists (their name taken from the location of their first meeting, Pugwash, Canada) formed in 1957 in response to a manifesto published in 1955 by Bertrand Russell and Albert Einstein. This network brought together scientists from around the world to discuss the "threat posed to civilization by the advent of thermonuclear weapons," which were vastly more destructive than fission weapons and produced more radioactive fallout.[27] In the United States and Europe, domestic advocacy groups also began to form, such as the National Committee for a Sane Nuclear Policy, the Committee for Non-Violent Action, and the Campaign for Nuclear Disarmament.[28] While Soviet antinuclear actions were motivated by Cold War international politics, these grassroots activists were motivated by a genuine and growing fear of nuclear war and an altruistic drive to prevent a global catastrophe.

Demonstration and Framing

Norm evolution theory highlights persuasion, framing, and shame as key mechanisms actors can use to promote an emerging norm, and these mechanisms can clearly be seen in the history of the emergence of the constraining nuclear norm. A tangible shift occurred when a critical mass of norm entrepreneurs began using persuasion and shame to foster the perception that nuclear weapons were different from conventional weapons and framing them in the context of poisonous chemical and biological weapons. The advent of thermonuclear weapons (also known as hydrogen bombs or H-bombs) in 1952 further fueled these motivations.[29] Soon the United States and the Soviet Union were conducting numerous atmospheric tests of these more powerful nuclear weapons, generating secondary fears about adverse health and environmental consequences due to radioactive fallout. These concerns reached a new level in July 1953, when a radioactive component associated with fallout from nuclear weapon tests, Strontrium-90, was detected in animals and milk.[30] In March 1954 this issue further galvanized public perceptions when the Japanese fishing boat *No. 5 Lucky Dragon* encountered radioactive fallout from a U.S. test of a fifteen-megaton hydrogen bomb, code-named Castle Bravo.[31] This led to panic regarding contaminated fish. These events and others inspired these activists—who were effectively serving as norm entrepreneurs to the extent they influenced perceptions and expectations of various actors—to focus on a more limited norm and restrictions against nuclear testing rather than a total norm against possession and/or use. Numerous activists in the peace movement, organized by groups such as the Committee for Non-Violent Action, held protests, circulated petitions, put out advertisements in newspapers and publications, and even violated the law by trespassing onto nuclear-related sites.[32] In the United States this led to a shift in public opinion, with a majority supporting a no-first-use policy for nuclear weapons.[33]

Reframing led to discussion regarding the moral acceptability of nuclear weapons and increased the emerging nuclear norms coherence and connection with key ideas and adjacent norms for chemical and biological weapons. Norm evolution theory indicates that these factors will accelerate norm emergence, and this certainly was the case with the emerging nuclear norm. This task was made somewhat easier by early U.S. moves to treat nuclear weapons as different from conventional explosive weapons. For example, President Truman put nuclear weapons into civilian control (the U.S. Atomic Energy Commission) rather than under military control and reserved sole discretion as to when to employ them.[34] This had never been done for any other type of weapon. Once the Soviet Union also possessed nuclear weapons and the fear of a global nuclear war began to marinate in the public consciousness, nuclear weapons' status as a special category of weapons was further solidified. President Eisenhower would partially reverse this special status by giving the military some launch authority, but the special status established by Truman's actions remained. The emergence of this new issue area shaping public opinion created a positive environment for norm emergence. The global experience of devastation and war in World War II and the budding Cold War coupled with the incredible new power of nuclear and eventually thermonuclear weapons fueled concerns and led nuclear norm entrepreneurs to engage a "global constituency" to address the threat of nuclear weapons.[35] The military redefined nuclear weapons as symbolic deterrent resources and not conventional armaments.[36] At times this new conception and symbolism of nuclear weapons—while constraining their use—made them enticing for nations seeking special status, such as the French.

Nuclear Weapon Norm Cascade

In the mid-1950s the emerging constraining nuclear norm began to reach a tipping point and a norm cascade began to occur. Key actors and international organizations, such as the Pugwash participants and UN entities such as the new Atomic Energy Commission, and various grassroots antinuclear groups actively promulgated information and helped shape and spread the new norm. These actors and organizations actively sought to categorize nuclear weapons as special (and thus stigmatize their use through coherence with existing WMD norms) and took advantage of the optimal timing (which is more apparent in hindsight) of the era as well as the norm's clear, specific, and universal claims.

The success of transnational norm entrepreneurs in stigmatizing nuclear weapon development (especially testing) and use (especially first use) created significant pressure that constrained international actors. The U.S. government actively sought to resist this pressure and use nuclear weapons to their tactical advantage. In 1954 the acting secretary of state sent a memo to the National Security Council identifying a need to reduce the "moral stigma" connected with nuclear weapons and Secretary of State Dulles regularly sought to break down the "false distinction" between conventional and nuclear munitions.[37] The U.S. military also actively sought to combat this stigma and treat nuclear weapons as just another weapon in their arsenal, promoting the competing permissive nuclear norm from 1945. The National Security Council's 162/2 document, developed by the military services and Joint Staff and approved in late 1953, identified the goal of treating nuclear weapons as "conventional" and "as available for use as other munitions."[38] In the multilateral realm in 1954, the United States and NATO adopted the policy of first use of tactical nuclear weapons in response to a Soviet and Warsaw Pact attack.[39] However, these efforts to combat the budding norm cascade associated with nuclear weapon development and use would not be successful as a critical mass of actors employing key mechanisms pressured states to support the norm for legitimacy and reputational reasons. For example, a 1955 National Intelligence Estimate concluded, "There is increased public pressure on governments to find some means of international disarmament, and especially some means of ensuring that nuclear weapons will not be used in war."[40]

Beyond protests, statements, and government policies, perhaps the best evidence that the emerging nuclear norm had reached a tipping point was the Korean War. During this war, a possessor of nuclear weapons (the United States) did not use them when facing nonnuclear adversaries (North Korea and China) despite realist power considerations dictating they should.

Actors and Motivations

The nuclear norm strengthened and reached a norm cascade in large part due to the fact that a significant number of actors were actively promulgating information and helping to shape and spread the new norm. This is consistent with norm evolution theory, which predicts that a norm cascade is more likely to occur when powerful actors (states, international organizations, and transnational networks) with the ability to create new institutions, laws, and expectations and incentives around the issue are involved. Norm evolution theory also speaks to why these key actors would be more likely to support and participate in a norm cascade. In the case of the nuclear norm, actors were motivated by legitimacy, reputation, and esteem. The debate over whether or not nuclear weapons were akin to conventional weapons led to a growing sense that states that did not draw some kind of special distinction for nuclear weapons lacked legitimacy. The grassroots antinuclear movement had succeeded in affecting public opinion, jeopardizing the reputation and esteem of states and actors that did not treat nuclear weapons as special and limit or constrain their use. This was also true in the domestic context in various democratic states where elected politicians were subjected to the public's growing antinuclear views. Therefore, reputation and esteem were key motivating factors. This is best seen in President Eisenhower's statement to Secretary Dulles that "as much as two-thirds of the world, and 50% of the U.S. opinion opposes" their policy of nuclear first use under massive retaliation.[41] A few months later he would sharpen his observation of the new normative reality when he told his staff, "The new thermonuclear weapons are tremendously powerful; however, they are not . . . as powerful as is world opinion today in obliging the United States to follow certain lines of policy."[42]

Beyond the mechanism of persuasion and diplomacy, actors used socialization through the delegitimation politics based on the creation of a new and "different" concept of WMD. The National Committee for Sane Nuclear Policy, the Committee for Non-Violent Action, and the Campaign for Nuclear Disarmament organized regular demonstrations (which garnered media coverage) that stressed the special destructive power of nuclear weapons.[43] The ongoing U.S. and Soviet thermonuclear weapon tests (and related incidents, such as the 1953 report of radioactive Strontrium-90 contamination) also demonstrated the special nature and danger of nuclear weapons that further helped socialize the new norm and helped it reach a cascade, particularly aspects related to testing and use, although the overarching constraining norm was bolstered.

Key Factors Associated with the Norm and the Environment

Certain environmental factors and factors associated with the norm itself enabled the nuclear norm to reach a tipping point. These factors are among those identified by norm evolution theory as important in making it more likely that a norm cascade will occur: optimal timing, coherence, and clear, specific, and universal claims. The optimal timing of the growing international media network and ease of communication through television, radio, and print enabled the norm and information related to its growth to spread more rapidly than otherwise possible. For example, the noble laureate Albert Schweitzer called for an end to nuclear testing in a series of widely heard radio addresses in 1957.[44] Television was becoming widespread, and television entertainment and news both regularly featured themes and demonstrations (fictional or real) of nuclear weapons that further accelerated awareness and socialization. This led to a boom in sales of bomb shelters and civil defense preparedness supplies in the United States; more important, it raised awareness of the nuclear threat and helped shape public support for the new norm. Because nuclear weapons were perceived as special and different, it was easier to promote a relatively simple norm that made universal claims on these weapons. Categorizing nuclear weapons as WMD tied them to existing norms for chemical and biological weapons, and thus the norm had a high degree of coherence with constraining norms for those other WMD modalities. Maria Rost Rublee examined this coherence with biological and chemical weapons and the special status afforded nuclear weapons as major factors that influenced how Japan, Egypt, Libya, Sweden, and Germany conceptualized and valued these weapons and ultimately chose not to pursue them.[45]

Nuclear Weapon Norm Internalization

Norm internalization, the final stage in the norm life cycle, can actually make the norm hard to discern because all debate and controversy surrounding it is resolved and discourse ends.[46] The international arms control and disarmament bureaucracy and the increasing regulation and legalization of armed conflict provided an increased number of organizational platforms and professional networks to institutionalize the norm. Ultimately, internalization of aspects of a nuclear norm governing usage occurred more rapidly and was easier to achieve than aspects governing development, proliferation, and disarmament.

The nuclear norm internalization stage began in the early 1960s as the competing view of nuclear weapons as another conventional-equivalent weapon faded away. The constraining nuclear norm—conceptualizing nuclear weapons as special and not acceptable for use except in extreme circumstances—began to be institutionalized through bilateral arms control efforts as well as the UN disarmament bureaucracy.[47] Tannenwald identifies the stalemate between the United States and the Soviet Union and the growing role of nuclear deterrence as part of this stalemate as a key factor in creating an environment favorable for internalization of the nuclear norm. She also points out that the rise of a block of nonaligned movement (NAM) countries in the UN General Assembly put additional pressure on the nuclear superpowers to adhere to the new nuclear norm and pledge no first use and ultimately seek arms control and disarmament efforts. Another key factor in the internalization of the nuclear norm was the antinuclear views of key U.S. politicians and policymakers, such as Secretary of Defense McNamara and U.S. Secretary of State Dean Rusk, who considered nuclear weapons unusable for moral reasons.[48] Their adherence to the view—expressed in the nuclear norm itself—that nuclear weapons were far from conventional and that their development and use should be subject to special constraints was likely a product, in part, of the efforts of norm entrepreneurs in the norm emergence stage. These factors helped internalize the nuclear norm—first in the area of the development and testing and then in disarmament and nonproliferation.

Professional Networks and the Arms Control Bureaucracy

Norm internalization began first with the aspect of the new nuclear norm that focused on the testing of nuclear weapons and was enabled by professional networks and the international arms control bureaucracy. Yield-producing tests of nuclear weapons played a pivotal role in developing nuclear weapons and refining their effectiveness, explosive capacity, and weight-to-yield ratios (essential for mating warheads with delivery systems). Beginning with the norm emergence state, concerns regarding the environmental impact of these tests as well as interest in slowing the development and proliferation of nuclear weapons served as catalysts for focusing on limiting and ultimately prohibiting these nuclear tests. This reached a new status when, in 1963, President John F. Kennedy signed the Limited Test Ban Treaty (LTBT) along with his Soviet counterparts.[49] The LTBT prohibited nuclear testing in the atmosphere, in outer space, and under water and currently has 108 signatories.[50] Of note, China did not sign the LTBT and in 1964 called it a "big fraud" intended to strengthen the nuclear monopoly of existing nuclear powers.[51] The institutionalization of the testing aspect of the nuclear norm was further bolstered in 1974 when President Nixon negotiated the Threshold Test Ban

Treaty (TTBT) limiting underground nuclear tests to a yield of 150 kilotons or less, roughly ten times the power of the U.S. bomb dropped on Hiroshima in 1945.[52] A few years later President Gerald Ford negotiated the Peaceful Nuclear Explosions Treaty (PNET), extending this kiloton limit to peaceful nuclear explosions, which were exempt from the TTBT. While these efforts represented progress in institutionalizing and codifying the norm, they were not fully and immediately adopted by the key actors: the United States and the Soviet Union. Only once the PNET was negotiated were both treaties submitted for ratification by the U.S. Senate.[53] However, due to concerns regarding the PNET and TTBT verification provisions, the United States and Soviet Union did not ratify the treaties (and consequently they did not enter into force) until 1990. However, this became more of a legal technicality in terms of the strength of the norm's internalization as both nations did observe the 150-kiloton limit while verification was pending.[54] Professional networks and the growing arms control bureaucracy (based on a great deal of legally sensitive issues such as treaties, conventions, and customary international law) buttressed the internalization of this norm—as norm evolution theory indicated it should.

Further evidence of the internalization of this norm occurred in 1991, when Secretary General Mikhail Gorbachev announced a unilateral moratorium on Soviet nuclear testing.[55] In 1992, with the end of the Cold War, President George H. W. Bush signed into law the fiscal year 1993 Energy and Water Development Appropriations Act, which enacted a unilateral U.S. moratorium on underground nuclear testing and on the introduction of new weapon designs into the U.S. nuclear stockpile.[56] This ended what had been the continual development of new nuclear weapons in the United States. Prior to this moratorium the United States had conducted 1,054 nuclear tests, and the collective international community had conducted roughly

2,000.[57] The moratorium, which was also observed by the Russian Federation, illustrates the negative impact the anarchic international environment can have on norm evolution. While the great power politics of the early Cold War helped incubate the new nuclear norm, it also held back its deep internalization in terms of a total ban on nuclear testing. As a hedge against potential threats, neither superpower was willing to fully abandon nuclear tests until the Berlin Wall fell. This power-balancing environmental factor—not previously identified in norm evolution theory—seems to inhibit deep norm internalization.

Deeper internalization of the testing aspect of the nuclear norm continued in the mid-1990s as momentum for a comprehensive ban on nuclear testing was building. Between January 1994 and August 1996 the Conference on Disarmament negotiated the Comprehensive Test Ban Treaty (CTBT), which was adopted by the UN General Assembly in September 1996 (bypassing Indian opposition).[58] The CTBT is the long-sought-after silver bullet when it comes to arms control related to nuclear testing and represents full codification of the testing aspect of the constraining nuclear norm. It requires member countries "not to carry out any nuclear weapon test explosion or any other nuclear explosion" and established a robust verification regime based on an International Monitoring System (IMS) of ground-based stations (collecting seismological, hydroacoustic, infrasound, and radionuclide information) and an International Data Center (IDC) to manage IMS-generated data. A process for on-site inspections upon request by any member state is also provided, pending an affirmative vote of at least thirty of the fifty-one members of the CTBT Executive Council.[59] The CTBT also creates a Comprehensive Test Ban Treaty Organization (CTBTO), although only a Preparatory Commission currently exists since the treaty is not in force. In order for the CTBT to enter into force, all states listed in Annex 2 of the treaty must ratify the CTBT.

Forty-four states are identified in Annex 2, thirty-five of which have ratified the treaty. The nine holdouts are North Korea, India, and Pakistan (all of whom have not signed the treaty) and China, Egypt, Indonesia, Iran, Israel, and the United States (all of whom have signed the treaty but not ratified it).[60] President Bill Clinton put forth the CTBT for Senate ratification in 1999, only to have the Senate vote 51–48 against ratification (sixty-seven votes are necessary per the Senate's constitutional advise and consent role).[61] The Senate rejected the treaty due to concerns that it was not verifiable, impaired the ability of the United States to maintain its nuclear stockpile, and provided limited nonproliferation benefit.[62] Many observers believe U.S. ratification of the CTBT would spur most or all of the other eight Annex 2 holdouts to ratify the treaty as it would bolster the motivation to conform.[63] Holdouts are given cover while the United States still refuses to ratify the treaty. Its failure in the United States could be due in part to the complexity associated with codification of the norm. However, the testing aspect of the constraining nuclear norm prevails in spite of these efforts and is largely internalized—with a significant minority of only India, Pakistan, and North Korea challenging the norm's dominance (three of nine nuclear weapon states)—and then only in rare and limited underground nuclear tests. Atmospheric nuclear testing is now unthinkable.

As internalization of the nuclear norm in regard to testing occurred, internalization moved forward in the area of disarmament and nonproliferation, also enabled by professional networks, lawyers, and the international arms control bureaucracy. This occurred internationally, beyond the bilateral arms control agreements between the United States and the Soviet Union, the Strategic Arms Limitation Talks (SALT) I (1972), SALT II (1979), and the Intermediate-Range Nuclear Forces treaty (1987). Soon after the enactment of the LTBT in 1963, negotiations began on the Nuclear Nonproliferation Treaty (NPT), which moved beyond testing and covered possession and disarmament. These negotiations lasted from 1965 to 1968 as a result of a variety of factors, but two in particular pushed the negotiations forward. The Chinese tested a nuclear device in October 1964 (stoking fears of proliferation built, in part, by norm entrepreneurs) and the UN Commission on Disarmament voted in June 1965 in favor of pursuing a treaty or convention on nuclear *nonproliferation*, a new term arising out of fear of an increase in the number of nuclear weapon states (horizontal proliferation) and in the number of bombs and warheads possessed by existing nuclear powers (vertical proliferation).[64] Due to the difficulty associated with the technology, nuclear weapons had not yet spread.[65] Under the auspices of negotiating a nonproliferation treaty that codified the nuclear norm, the Soviet Union and the United States offered different initial proposals, largely reflecting disagreement over whether or not U.S. nuclear weapons committed to NATO constituted unacceptable proliferation.[66] That issue was ultimately resolved, and in June 1968 the Commission on Disarmament offered a resolution to the UN General Assembly in support of the draft NPT, which was overwhelmingly adopted. The treaty opened for signature in July and went into force in March 1970.[67]

The NPT reflected the internalization and codification of various aspects of the nuclear norm. First, the NPT requires the five recognized nuclear weapon powers (the United States, the Soviet Union, the United Kingdom, France, and China) to not transfer or proliferate nuclear weapons or technology to other nations. Second, Article VI requires all parties to "pursue negotiations in good faith on effective measures relating to cessation of the nuclear arms race at an early date and to nuclear disarmament, and on a treaty on general and complete disarmament under strict and effective international control."[68] The nonnuclear weapon powers (also known as nonnuclear weapon states or NNWS) are required to not seek to develop or obtain nuclear weapons; this is enforced by the International Atomic Energy Agency (IAEA), using various safeguards to detect diversions of nuclear material from efforts to peacefully acquire nuclear energy. As the NPT opened for signature, the nuclear powers sought to allay the concerns of the NNWSs by pledging to provide assistance to any state subject to or threatened with nuclear attack.[69] It was a few years before some key countries, such as the Federal Republic of Germany and Japan, ratified the treaty, but the overwhelming majority of nations ultimately did so. They were motivated in part by the goal of conformity with the nuclear norm codified in the NPT. As of October 2013, 190 nations ascribe to the NPT, and only India, Israel, Pakistan, and South Sudan have not signed.[70] North Korea was a party to the treaty but announced its controversial withdrawal in January 2003. The legal and professional IAEA bureaucracy created by the NPT led to greater internalization of the nuclear norm through professional networks and habit-based recurring interaction—consistent with norm evolution theory. The institutionalization of the nuclear norm through the NPT and other means has been so effective as to render proliferation estimates from the early nuclear era completely off base. In March 1963 President Kennedy publicly estimated that fifteen to twenty-five states might obtain nuclear weapons by the 1970s.[71] As summarized by Rublee, the NPT "dramatically changed the cost-benefit equation for nuclear weapons" by imposing additional costs for violating the norm.[72]

Domestic and Regional Issues

The constraining nuclear norm regarding proliferation, enshrined in the NPT, is not without challenge. While norm evolution theory indicates domestic turmoil can help motivate actors to support a norm's internalization, the nuclear experience shows it can also spur efforts to buck the norm. In 1998 both India and Pakistan openly tested nuclear weapons as part of their efforts to acquire a nuclear deterrent force and in response to internal turmoil and tension over their competition and conflict. These nuclear tests shocked the world; however, neither nation was an NPT signatory, and following the test India announced it was adopting a no-first-use policy and that it would never use nuclear weapons against nonnuclear weapon states.[73] In any case, India's test largely symbolic as it lacked suitable delivery vehicles or a nuclear employments strategy.[74] Its motivation was likely similar to France's in the 1950s: to achieve great power status, an unfortunate secondary effect of the special unconventional status afforded nuclear weapons by the nuclear norm. Pakistan has not made a similar no-first-use pledge, but it has broadcast nuclear restraint and refrained from an aggressive nuclear-use policy, demonstrating an adherence to some principles of the nuclear norm (particularly those regarding use), if not the NPT. However, the godfather of the Pakistani nuclear program, Abdul Qadeer Khan, has exhibited no constraint regarding the proliferation aspect of the nuclear norm as he illicitly spread nuclear technology on the black market.[75] The fact that Khan had to do so covertly indicates that the proliferation aspect of the constraining nuclear norm still had some force, and the extent to which the Pakistani military or government sanctions Khan's efforts is unknown.

This adherence to the usage aspect of the nuclear norm and the effort to hide violations of the proliferation aspect are likely products of deliberate norm cultivation using motivating factors such as conformity and institutionalization. Thomas Schelling discusses the work of international dialogue (facilitated through organizations like the Aspen Institute and the Institute for Strategic Studies in London) in spreading and promoting the norm against nuclear use to new members of the nuclear club, such as India and Pakistan.[76] North Korea has been a particularly disconcerting outlier when it comes to both the NPT and the nuclear norm in general, perhaps because it is not motivated by factors such as conformity. North Korea has sporadically tested nuclear weapons over the past decade, and its withdrawal from the NPT remains legally controversial.[77] In recent years some NNWSs, especially those from the NAM block, have come to view the Article VI disarmament requirement as more of a vague negotiating element required to get them to agree to the treaty and the preferred status the NPT provides to existing nuclear powers. These activities lead to reasonable questioning as to whether the nuclear norm embodied in the NPT and the treaties related to nuclear weapon testing truly reflect the norm's internalization.

Leaders of Key Professions

Despite the setbacks of India, Pakistan, and North Korea, there are signs of progress for the nuclear norm's internalization. Norm evolution theory indicates that the involvement of particularly relevant professions is key to internalizing the norm. In addition to the scientists who provided pivotal support in strengthening the testing aspect of the constraining nuclear norm, the national security professions are the most relevant. Recent developments by senior statesmen have also helped renew and reinvigorate efforts to more deeply internalize the overall nuclear norm, particularly its disarmament aspect. Over the past decade there has been a resurgent interest in total nuclear disarmament, showing that the nuclear norm may in fact be so deeply internalized that the idea of total nuclear disarmament can again be revisited. In January 2007 four senior U.S. national security officials—George Schultz, Henry Kissinger, Bill Perry, and Sam Nunn, referred to by some as the "Four Horsemen"—published an op-ed in the *Wall Street Journal* calling for a world free of nuclear weapons.[78] This was not the first call for nuclear disarmament; similar efforts have existed since the advent of nuclear weapons.[79] Following the end of the Cold War many believed that nuclear weapons were no longer required; among these were sixty high-ranking military officials from numerous countries who issued a statement in 1996 calling for complete nuclear disarmament.[80] The 2007 *Wall Street Journal* op-ed was particularly noteworthy because it came from a bipartisan group of former cabinet-level statesmen in the national security profession who did not have reputations for Pollyannaish thinking and were known as realists during their tenure in government. These individuals could be considered key actors and norm entrepreneurs (in general as well as within their profession), per norm evolution theory. Suddenly the nuclear disarmament movement had more credibility than ever before.

Other senior leaders from the United States and the international community as well as a variety of think tanks proceeded to follow the statesmen's call and come together in support of this goal in a fresh effort called Global Zero. On April 5, 2009, this movement received the highest level of support by any U.S. administration when President Barack Obama gave a speech in Prague declaring the existence of thousands of nuclear weapons as "the most dangerous legacy of the Cold War" and proceeded to commit the United States to a renewed effort to seek a world without nuclear weapons.[81] Additional senior U.S. leaders have signed on to the Global Zero effort, including a number of other cabinet-level officials from the past six administrations.[82] The Stimson Center, a nonpartisan national security think tank and participant in the Global Zero initiative, has published a series of papers under the title *Unblocking the Road to Zero* that examine states' views regarding the utility of their nuclear weapons and the conditions necessary for nuclear disarmament. Nations covered include allies such as the United Kingdom, France, and Israel as well as India, China, and Pakistan.[83] A review of these national perspectives does not raise hopes that security conditions can be achieved that would allow them to reduce to zero. For example, Bruno Tertrais, a French defense academic, identifies the elimination of all major threats against France's broadly defined vital interests and those of its European allies as a precondition for abandoning its roughly 350 nuclear weapons.[84] Tertrais identifies the global elimination of chemical and biological weapons and the end of the "offense-defense" arms race as further preconditions. Brig. Gen. Shlomo Brom (retired) and Brig. Gen. Feroz Khan (retired) of Israel and Pakistan, respectively, highlight their countries' views that their nuclear weapons are essential to offset their neighboring adversaries' greater size and perceived conventional military capabilities.[85] While this highlights the view of some key actors in the national security profession as well as the long road ahead for total internalization of the nuclear norm, the fact that the international and professional dialogue has moved in this direction appears to be a significant step in further institutionalizing and motivating conformity with the nuclear norm.

While nuclear norms regarding testing, proliferation, and disarmament are to varying extents codified in international agreements, no such comprehensive codification has occurred regarding the more basic (and earliest to emerge) norm regarding nuclear weapon use (including first use). T. V. Paul, a thought leader on the topic of nuclear norms, approaches the issue of the nuclear norm through a practical, realist lens and concludes that there are no explicit or legally binding prohibitions against the use of nuclear weapons (the NPT does not address the issue of use).[86] Thus he considers the nuclear usage norm a "tradition" of nonuse rather than a more constraining or binding "taboo." He argues that this normative tradition was developed and influenced by the "logic of consequences," a realist and rationalist argument regarding self-interest and the negative effects nuclear weapon users would experience, and the "logic of appropriateness." But whether a tradition or a binding legal construct, the aspect of the nuclear norm regarding use is largely internalized in the sense that it has achieved taken-for-granted status and remains unchallenged. This is not to say all nuclear weapon states have an overt no-first-use policy; rather they simply have not used nuclear weapons. Recognition of the internalization of the constraining nuclear norm regarding use was acknowledged even by one of the strongest proponents of increased planning for using nuclear weapons and "winning" a nuclear war: Herman Kahn. In 1966 Kahn wrote, "Prevailing attitudes towards 'first use' of nuclear weapons have changed considerably. The United States will probably adopt a virtual 'no-first-use' policy—perhaps without making a deliberate decision to do so."[87]

Whereas the usage aspect of the nuclear norm was not formally codified through a mechanism such as the NPT, it was in some cases regionally codified by the establishment of "nuclear-weapon-free zones." These limited possessing, testing, deployment, and use of nuclear weapons in certain regions. The first such zone was proposed in mid-1962 by Brazil and then codified in the Treaty of Tlatelolco in February 1967.[88] Later, four additional regional agreements were established to create additional zones in areas in the South Pacific, Southeast Asia, Central Asia, and Africa.[89] These regional denuclearization treaties further engrained the internalization of the nuclear norm. By the 1980s the nuclear norm was so deeply internalized that nuclear war plans and strategies that were plausible in the 1950s were considered far less tenable.[90] This is not to say potent nuclear capabilities no longer exist, just that expectations regarding their use have made such use less likely. While not legally codified in a broad way, the nonuse aspect of the nuclear norm has "served well thus far to achieve what is the declared goal of disarmament: preventing nuclear war."[91] Other factors, such as deterrence through the threat of nuclear retaliation, certainly also contributed to this outcome, but norms have played an enduring role.

Summary

The emergence, cascade, and internalization of constraining norms for nuclear weapons were consistent with many of the various hypotheses from norm evolution theory. While the initial environment was not favorable for the nuclear norm due to the legacy of World War II and strategic bombing, early Cold War dynamics and international organizations and platforms helped serve as incubators for the growth of norm entrepreneurs. Persuasion, especially through framing based on the proposed norm's coherence with related norms for poison (chemical and biological) weapons, exploited increasingly globalized communication and professional networks to expedite the norm's spread, and by the early 1960s a tipping point had been reached. This framing also firmly established nuclear weapons as a unique threat and different from conventional armaments.[92] The slow spread of nuclear weapons, due to technical challenges, also allowed time for entrepreneurs to organize and to challenge previous understandings and for the norm to emerge and reach a cascade. For a time the United States was the sole nation in possession of nuclear weapons and thus had a decisive role in establishing behavior. While key U.S. leaders wished that the constraining norm did not exist, they inadvertently furthered the constraining nuclear norm and stigma associated with nuclear weapons. U.S. restraint in exploiting its temporary nuclear dominance through nuclear use as well as its pursuit of strategic advantage by limiting proliferation and even seeking to eliminate state possession altogether through the Baruch Plan allowed space for a constraining nuclear norm to emerge. This window was successfully exploited by norm entrepreneurs and organizations, some of which were encouraged by the Soviet Union's demagoguery of the U.S. nuclear monopoly as part of the power balancing of the early Cold War. Power balancing, while inadvertently fostering norm emergence, can also inhibit deep norm internalization, as was seen with the delay in final codification of the testing aspect of the nuclear norm in the CTBT.

Some of this success in the nuclear norm's emergence, cascade, and internalization was clearly due to contingency and chance. This is evidenced in the U.S. restraint in nuclear use as well as the fact that optimal timing occurred as the weapons themselves spread slowly due to technical challenges. One effective mechanism used by norm entrepreneurs was the framing of nuclear weapons as special unconventional weapons, known as WMD. Over time internalization of aspects of the nuclear norm governing usage occurred more rapidly and was effectively easier to achieve than aspects governing development, proliferation, and disarmament. This may have been in part due to the fact that monitoring compliance with the norm for use is much simpler than doing so for development, proliferation, and disarmament. There is little doubt when a nuclear weapon is used but much uncertainty regarding the existence and scope of efforts to develop nuclear weapons, as is seen today with the uncertainty regarding Iran's nuclear program and intentions in spite of the regime's claims that its program is peaceful and that nuclear weapons violate Islamic teaching. Table 7 summarizes which hypotheses from

Table 7. Hypotheses of norm evolution theory applied to nuclear weapons

HYPOTHESES	NUCLEAR WEAPONS
NORM EMERGENCE	
Significant presence of norm entrepreneurs/norm leaders with organizational platforms	XX
Key actors motivated altruism, empathy, and ideational commitment	X
Actors employ persuasion, framing, memes, shame	XX
Domestic political concerns	X
States seeking international legitimation	
State distribution of power, prevailing levels of technology, availability of natural and human resources, etc. favoring the norm	X
Large-scale turnover of decision makers	
Recognized failure of norms from a previous norm generation	
Emergence of a new issue area where prevailing norms not well established	X
Norm prominence	
Norm coherence—connections to key ideas, adjacency, and grafting	XX
NORM CASCADE	
Key actors are involved, specifically states, international organizations, and transnational networks	XX
Key actors motivated by legitimacy, reputation, esteem	XX
Actors employ persuasion and diplomacy	X
Actors employ institutionalization	XX
Actors employ socialization	X
Actors employ demonstration	X
Optimal timing	XX
Norm clear and specific	X
Norm makes universal claims	X
Norm coherence	X

Table 7. (*continued*)

HYPOTHESES	NUCLEAR WEAPONS
NORM INTERNALIZATION	
Key actors are involved, specifically law and lawyers, professions, and bureaucracy	XX
Actors motivated by conformity or domestic turmoil	XX
Actors employ habit and institutionalization	XX
Existence of professional networks	XX
Norm clear and specific	X
Norm makes universal claims	X
Norm coherence	X

More prevalent variables are indicated by XX rather than X.

norm evolution theory were present in the history of the evolution of norms for nuclear weapons.

While the hypotheses of norm evolution theory are consistent with the experience of the constraining nuclear norm's life cycle, they are too numerous and indeterminate to be particularly useful for offering predictions for the future emergence of constraining norms for other emerging-technology weapons. This chapter has identified a few areas where the nuclear experience offers some particularly valuable additional lessons to compliment and further unpack the general hypothesis of norm evolution theory. These include the fact that characterizing the weapon type as unconventional or otherwise granting it a special status can accelerate norm adoption and ultimately achievement of a norm cascade. Unitary dominance of a single actor with the particular weapon type can give that actor significant influence in influence norm emergence for that weapon type. Similarly, delays in a weapon type's proliferation (often due to technological barriers) can create added time for a constraining norm to emerge. The international arms control and disarmament bureaucracy and the increasing regulation and legalization of armed conflict provide an increased number of organizational platforms and networks to spread the norm and more rapidly achieve a norm cascade. Finally, the internalization of aspects of a norm governing usage occurs more rapidly and is easier to achieve than aspects governing development, proliferation, and disarmament.

5 Norm Evolution Theory for Emerging-Technology Weapons

> You can't say civilization don't advance. . . . In every war they kill you in a new way.
>
> Will Rogers

In the first century AD, the Roman statesman Frontinus said, "[The] engines of war have long since reached their limits, and I see no further hope of any improvement in the art."[1] His wildly inaccurate prediction illustrates the wisdom of Lao Tzu's axiom that those who have knowledge don't predict, and those who predict don't have knowledge.[2] Thus it is with humble awareness of the limits and challenges of prediction that I seek in this chapter to offer a framework to do so, particularly for the volatile and changing arena of emerging-technology weapons. The previous three chapters applied general norm evolution theory to three historic case studies: CBW, strategic bombing, and nuclear weapons. The general hypotheses of norm evolution theory were for the most part validated in these case studies. However, norm evolution theory largely consists of a laundry list of hypotheses and is not developed specifically for nor tailored to norms for emerging-technology weapons and warfare. As such, it is not optimized for making predictions of norm evolution for emerging-technology weapons, such as those for cyber weapons. This chapter seeks to offer a more refined version of norm evolution theory tailored for emerging-technology weapons based on the preceding case studies.

Norm Evolution Theory Applied to Case Studies

Norm evolution theory offers specific hypotheses regarding the actors, actor motives, and important mechanisms and factors supporting norm development during all three stages of the norm life cycle. The historic case studies of CBW, strategic bombing, and nuclear weapons found the theory's hypotheses consistent with the norm evolution experience for these three modes of warfare. See table 8 for a summary of norm evolution theory's hypotheses and their presence or absence in each of the case studies.

While each of the three norms of the case studies is in the third stage of the norm life cycle and well established, their development has followed different paths, with different fits and starts and more or less important variables (as indicated by the variation in the single or double Xs in table 8). Further, they are not all equally internalized today. For example, both the CBW and nuclear norms are relatively well codified in the Geneva Protocol, BWC, CWC, and NPT. No similarly focused treaty exists for strategic bombing, and significant violations of the norm have occurred in Syria in 2012 and 2013 (leading to 4,300 deaths) without comparable international outcry for Syria's much more limited violation of the CBW norm.[3] Not surprisingly strategic bombing has a "score" of 6/2 (X/XX) for the internalization stage compared to a 7/5 and 7/4 score for CBW and nuclear weapons, respectively. This may be a result of the lack of

Table 8. Hypotheses of norm evolution theory applied to case studies

HYPOTHESES	CBW	STRATEGIC BOMBING	NUCLEAR WEAPONS
NORM EMERGENCE			
Significant presence of norm entrepreneurs/norm leaders with organizational platforms	X	XX	XX
Key actors motivated altruism, empathy, and ideational commitment	X	XX	X
Actors employ persuasion, framing, memes, shame	X	XX	XX
Domestic political concerns	X	X	X
States seeking international legitimation	X	X	
State distribution of power, prevailing levels of technology, availability of natural and human resources, etc. favoring the norm	X	XX	X
Large-scale turnover of decision makers			
Recognized failure of norms from a previous norm generation		XX	
Emergence of a new issue area where prevailing norms not well established		X	X
Norm prominence	X	X	
Norm coherence—connections to key ideas, adjacency, and grafting	XX	XX	XX
NORM CASCADE			
Key actors are involved, specifically states, international organizations, and transnational networks	X	X	XX
Key actors motivated by legitimacy, reputation, esteem	XX	XX	XX
Actors employ persuasion and diplomacy	X	X	X

Table 8. (*continued*)

HYPOTHESES	CBW	STRATEGIC BOMBING	NUCLEAR WEAPONS
Actors employ institutionalization	XX	XX	XX
Actors employ socialization	XX	X	X
Actors employ demonstration	X	X	X
Optimal timing	XX	XX	XX
Norm clear and specific	X	X	X
Norm makes universal claims*	X	X	X
Norm coherence	XX	X	X
NORM INTERNALIZATION			
Key actors are involved, specifically law and lawyers, professions, and bureaucracy	XX	XX	XX
Actors motivated by conformity or domestic turmoil	XX	XX	XX
Actors employ habit and institutionalization	XX	X	XX
Existence of professional networks	XX		XX
Norm clear and specific	X	X	X
Norm makes universal claims*	X	X	X
Norm coherence	XX	X	X

*Universal claims may still be selective in the sense that they apply only to "civilized" nations.
More prevalent variables are indicated by XX rather than X.

comparable professional networks and advocates as well as indicative of the lack of norm leaders (states) perceiving deep internalization in their interest. Professional networks, such as the chemical industry and the nuclear bureaucracy under the IAEA, are clearly a particularly important variable for deeper internalization. This alludes to the role of congruent public and private support for the norm, an important secondary hypothesis.

That said, with the exception of norm evolution theory's hypothesis that the large-scale turnover of decision makers will help enable norm emergence, elements from each hypothesis were present in one or more of the three case studies, with many in all three. Therefore, norm evolution theory remains a generally valid tool for analyzing the future prospects of norms for emerging-technology weapons, including cyber warfare. However, these hypotheses are too numerous and broad and not sufficiently discriminating to allow for detailed predictions for future norm development for emerging-technology weapons. Further, general norm evolution theory does not identify the primary independent variable leading to norm emergence and development, perhaps because it was originally designed for a wide range of purposes and not specific to international behavior and warfare. As such, it is necessary to refine and tailor norm evolution theory to make it more useful for this purpose.

Norm Evolution for Emerging-Technology Weapons

Norm evolution theory was not developed specifically to apply to norms for weapons and war. Many of the examples involve the development of norms for human and civil rights. Accordingly, examining how norm evolution theory, informed by the preceding case studies, specifically applies to norms for emerging-technology weapons is necessary to offer predictions. This examination allows for the identification of the primary independent variable that explains why constraining norms sprout and grow for emerging-technology weapons. It also offers the development of some specific secondary or modifier variables and related expectations or hypotheses concerning what one may anticipate for norm emergence and development for these weapons.

Emerging-technology weapons are weapons based on new technology or a novel employment of older technologies to achieve certain effects. Given that technology is constantly advancing, weapons that initially fall into this category will eventually no longer be considered emergent. For example, the gunpowder-based weapons that began to spread in fourteenth-century Europe would clearly be classified as emerging-technology weapons in that century and perhaps in the fifteenth century, but eventually these weapons were no longer novel and became fairly ubiquitous.[4] The same may be said for medieval longbow technology and strategic bombing and chemical weapons in the late nineteenth century up to World War II. Nuclear and biological weapons may be considered emerging-technology weapons during World War II and immediately after. Today cyber weapons that are used to conduct CNA are emerging-technology weapons. Cyber weapons are certainly not the only current contender for this label, as other novel weapons, such as autonomous weapon systems, high-energy lasers, and microwave-based active denial systems, may also be labeled emerging-technology weapons.[5]

Primary Hypothesis

When applied specifically to this category of emerging-technology weapons and informed by the case studies, norm evolution theory suggests that the principal independent variable that causes the emergence and development of constraining norms for emerging-technology weapons is the perception among powerful or relevant states that such norms are in their national self-interest. That is, a direct or indirect alignment of national self-interest with a constraining norm leads to norm emergence, and the extent to which it is aligned with key or powerful states' perception of self-interest will determine how rapidly and effectively the norm emerges. Self-interest and power can be broader than just material factors, as the international relations theorist Hans Morgenthau points out: "Power may comprise anything that establishes and maintains the control of man over man. Thus power covers all social relationships which serve that end, from physical violence to the most subtle psychological ties by which one mind controls another."[6] Thus a perception of self-interest can be due to a conscious "logic of consequences" calculation, discussed by Paul in the context of nuclear norms and often based on the realist perception that states are threatened by uncontrolled diffusion and use of the weapons. Alternatively states' perception of self-interest can be due to a "logic of appropriateness" determination based on the candidate norm's coherence with other norms the state embraces rather than a conscious state decision of material self-interest. As an example of the former, look no further than the Russian czar's interest and pursuit of a constraining norm for both CBW and strategic bombing at the First Hague Conference.[7] Russia was largely motivated by its own failures to develop these new weapons and an interest in constraining its adversaries. In the case of the CBW norm, this was in addition to the prospective norm's coherence with the poison taboo and the "logic of appropriateness." As another example of the role of state interest, Britain opposed the nascent strategic bombing norm because it saw aerial weapons as a potential tool to offset its army's numerical disadvantage when compared to France and Germany.[8] The role of national self-interest as the primary independent variable leading to norm emergence also helps explain why, when challenged with violations of a young and not-yet-internalized norm, a state is quick to abandon the norm and pursue its material interest by using the previously constrained emerging-technology weapon, as was seen with both CBW and strategic bombing in World War I and strategic bombing in World War II.

While direct national self-interest is the primary independent variable that explains why constraining norms for emerging-technology weapons initially emerge and develop, there is also a dynamic or systems theory element, which reflects that sometimes norm emergence is not the product of a single decision or set of decisions but is based on chance and the complex interaction of the overall system. The three historical case studies make it clear that chance, particularly the fortuitous and indirect alignment of a norm with a powerful and self-interested state actor, plays a major role in the early success of a norm for emerging-technology weapons. The Soviet efforts early in the nuclear era were intended to delegitimize the United States and strengthen the Soviet position while they aggressively pursued their own nuclear weapons. Instead, these efforts helped generate incubators for the constraining nuclear norm and spawned interest in the early antinuclear weapons movement. The Soviet-led World Peace Council's 1950 Stockholm Peace Appeal called for a total ban on nuclear weapons and involved a huge public outreach effort.[9]

In addition to this primary cause for norm emergence, norm evolution theory for emerging-technology weapons suggests that secondary variables, such as certain actors, actor motives, and important mechanisms and factors, support norm emergence and growth during each stage of the norm life cycle. While these hypotheses are identified for each of the three phases of the norm life cycle, they can apply in the additional stages. This refined theory of norm evolution for emerging-technology weapons is a more useful framework for predicting the likelihood of norm development for this category of weapons and warfare.

Secondary Hypotheses Concerning Norm Emergence

Hypothesis 1. Coherence and grafting with existing norms will play a key role in the early foundation of the norm for the emerging-technology weapon. Norm evolution theory identifies coherence with existing norms and norm entrepreneurs' deliberate grafting onto prominent existing norms as elements in a norm's success. This appears to be particularly true for emerging-technology weapons, and the initial expectations of behavior for the use, development, and proliferation of the new weapon will be directly tied to the dominant normative fabric on which its associated norms are (at least initially) affixed. This was seen with the coherence of constraining norms for chemical weapons with ancient norms and taboos regarding poison weapons and with the coherence of initial permissive nuclear norms with permissive norms for strategic bombing (although the constraining nuclear norm was ultimately able to overcome this coherence with the permissive nuclear norm).

Hypothesis 2. Permanently establishing a norm before the weapon exists or is fully capable or widespread will enhance a norm's chance of emergence and eventual cascade. Reaching an international consensus and then institutionalizing it before a weapon exists or is mature or widespread helps curtail competing permissive norms and establishes a firm foundation for the norm to achieve a tipping point and ultimately internalization. This was seen with the establishment of a prohibition on modern chemical weapons at the 1899 Hague Conference, which permanently banned the use of asphyxiating shells, although no such weapons had yet been invented.[10] While this was not enough to strengthen the norm to the point where it prevented CW use in World War I, it did help the norm develop. Additionally it appears that in order to achieve the benefit of this preemptive prohibition it needs to be permanent. For example, the prohibition on strategic bombing (indeed, any form of aerial bombardment) at the same 1899 conference was for a limited duration—five years—and when it expired and the fixed-wing aircraft and other tools for aerial bombing had been invented and were spreading, the development of the constraining norm suffered.

Hypothesis 3. There will be multiple challenges with undemonstrated emerging-technology weapons. First, because these weapons are not demonstrated, they will not be well understood and there will be differing perspectives on their potential capability and effects. This makes state calculations of self-interest—the key variable for norm evolution—more complex. Competing predictions on the impact of strategic bombing impeded a norm cascade for the constraining norm based on the view that strategic bombing would make war more costly. Specifically, in Douhet's 1921 book *Command of the Air* (and others), the idea was promoted that strategic bombing would actually make future wars shorter, with less overall noncombatant harm and damage.[11] Without a real demonstration or test of these competing predictions, the development of constraining norms was impeded. Inflation of the threat of the possible effects of the weapon can inadvertently strengthen norm emergence for a constraining norm. This inflated view of the threat is often caused by the private sector through industry and lobbying groups. As is to be expected, certain elements of private industry and public organizations will be disadvantaged by constraining norms for certain types of weapons. For largely unproven and novel emerging-technology weapons, constraining norms may even be viewed as a threat by public entities such as the military units or elements responsible for pursuing the new weapon. During the Washington Treaty and Geneva Protocol debates, the U.S. Army Chemical Warfare Service partnered with the American Chemical Society to make "totally irresponsible . . . exaggerations of new weapons developments" in order to play up the military utility of chemical weapons. This generated institutional support and business, as well as opposition to normative constraints and agreements.[12] These claims created an unrealistic and disproportionate fear of chemical weapons relative to their actual capabilities, which ultimately helped foster the constraining norm. Additionally, the lack of demonstration and experience with emerging-technology weapons can lead to inadvertent escalation due to a lack of clarity regarding weapon employment. This was seen with strategic bombing during the early days of World War I. In 1915 the Germans, French, and British perceived each other's aerial attacks as indiscriminate (and thus violating the nascent strategic bombing norm) even when the nonmilitary casualties were in fact inadvertent and due to the technological inability at the time to engage in precision bombing.[13] This led to further norm erosion due to retaliatory violations of the norm and inadvertent escalation.

Hypothesis 4. Connections with the idea that there is no defense against the weapon fuels interest in a constraining norm but also limits the effectiveness of reciprocal agreements and can lead to weapon proliferation. If states and their domestic audiences perceive the emerging-technology weapon as one that cannot be defeated with defenses, they will be more likely to perceive a constraining norm in their self-interest (the key variable for norm emergence) as it is one of the few options available to contain the threat. The view that there were no defenses against nuclear weapons helped motivate the United States to pursue international control of nuclear weapons under the Baruch Plan in 1946.[14] This contributed to the emergence of the international norm constraining nuclear weapon use. Interest in norms to constrain strategic bombing was similarly fueled by the idea that bombers could not be sufficiently defended against and would always reach their target.[15] This ultimately led to the French government's Bourgeois Plan to establish a multinational peacekeeping force, including the only bombing force in existence, under the League of Nations.[16] While this view of the futility of defenses fuels interest in constraining norms, it is not without risk.

It can lead to convention-dependent norms, which, when violated, result in intense domestic pressure for retaliatory violations of the norm. This was seen with the usage of chemical weapons in World War I and strategic bombing in World War I and II.

Hypothesis 5. Weapon proliferation and adoption will play a role in norm emergence by, among other things, influencing states' perception of self-interest. First, if there is a period when a single actor has unitary dominance with the particular emerging-technology weapon, it has increased influence as to how norms emerge for that weapon type. This was seen when the United States had an exclusive monopoly on nuclear weapons and, following the Hiroshima and Nagasaki bombings, resulted in significant nuclear restraint. This helped establish an operative precedent of nuclear nonuse and allowed time and space for the constraining norm to emerge. Second, a delay in an emerging-technology weapon's proliferation (often due to technological barriers) can create added time for a constraining norm to emerge. Technical hurdles associated with developing nuclear weapons prevented their broad diffusion (and potential temptations to violate the nascent constraining norm on nuclear weapon usage) even after the U.S. nuclear monopoly ended. Third, and perhaps most important, varied rates of adoption of the new technology will result in varied state interests in the norm and make the convergence of self-interest more difficult. Whether a state possesses or is successfully developing the emerging-technology weapon or the state views it as a threat to its existing military capabilities will influence its support or opposition to the constraining norm. Russia was the main norm leader for the constraining strategic bombing norm at the 1899 Hague Conference, motivated in part by the recent failure of its program to develop a dirigible airship.[17] Meanwhile the French and Germans were succeeding in their efforts to develop aerial weapons and thus were less receptive to the norm. Britain, while not suffering from the same technological failures as the Russians, supported the constraining norm because it viewed balloons as a potential tool to offset its army's numerical disadvantage.[18]

Secondary Hypotheses Concerning Norm Cascade

Hypothesis 1. Improvements in technology that address previous challenges in adhering to a constraining norm can rapidly lead to a norm cascade. At times the limits of the emerging-technology weapons will impair a particular constraining norm's success. For example, early bombing technology made the practical application of the constraining strategic bombing norm difficult as discriminate bombing was not necessarily possible, even if desired. However, the advent of precision-guided munitions in the years following the Korean War enabled a level of accuracy whereby bombing nonmilitary targets (however defined) while avoiding significant inadvertent collateral damage could be achieved. This technological breakthrough helped the reemerging strategic bombing norm succeed in achieving a norm cascade and ultimately internalization in some places. Similar technological breakthroughs can enable other emerging-technology weapons to overcome initial barriers to norm success.

Hypothesis 2. Characterizing the weapon type as "unconventional" or otherwise granting it a special status can accelerate norm adoption and ultimately the achievement of a norm cascade. By definition, emerging-technology weapons do not fit cleanly into existing conventional categories. The success of norm entrepreneurs in categorizing the weapons as special helps single them out for and accelerate norm development. This was seen with President Truman's decision to place nuclear weapons under civilian rather than military control and reserve sole discretion as to when to employ them—something that had never been done with any other type of weapon.[19] This is also seen with the unconventional category of WMD, including biological and chemical weapons, which international treaties declare to be "repugnant to the conscience of mankind."[20] However, such special categorization also runs the risk of incentivizing its proliferation by making it exotic, prestigious, or a symbol associated with modernity.[21] It was this symbolism associated with nuclear weapons that some sought to ascribe to other WMD modalities, as evidenced in 1988 when the speaker of the Iranian Parliament, Hashemi Rafsanjani, referred to biological and chemical weapons as "the poor man's atomic bomb."[22] If this type of association develops early for an emerging-technology weapon before constraining norms develop, the benefits of such an association could be negated.

Hypothesis 3. Public demonstrations of the weapon type, enabled by real-time media, can influence public opinion and spread the constraining norm. The lack of demonstration and understanding associated with an emerging-technology weapon can make norm development more difficult by impairing states' calculations of self-interest. However, once such demonstrations occur and are widely observed, especially with the modern 24/7 real-time media, they can help to quickly shape public perspectives on the weapon and help the constraining norm achieve a tipping point. For example, the reemerging strategic bombing norm received a large boost in support through media criticism of the U.S. bombing campaign in Vietnam. The press called the Linebacker II bombing campaign the "most savage and senseless act of war ever visited . . . by one sovereign people on another," and visual footage helped motivate more norm entrepreneurs and leaders through emotional evocation of altruism and empathy.[23] This modern media presence can be considered an element of optimal timing identified by norm evolution theory and will help ensure that the demonstration of emerging-technology weapons plays a role in norm development.

Hypothesis 4. The international arms control and disarmament bureaucracy and the increasing regulation and legalization of armed conflict provide an increased number of organizational platforms and networks to spread the norm and more rapidly achieve a norm cascade. Since the last few years of the nineteenth century the international community has created an increasing number of venues and forums to discuss, promulgate, and codify international norms. This was first seen with chemical weapons and aerial bombing at the Hague Convention in 1899 and then in 1907 and later the Geneva Convention, Washington Naval Conference, League of Nations, UN Commission on Disarmament, Nuclear Nonproliferation Treaty and International Atomic Energy Commission, and the OPCW. The increasing development and discussion of international law and acceptable international behavior ensure that norms for any emerging-technology weapons will have a multitude of organizational platforms and networks through which to spread and grow.

Secondary Hypotheses Concerning Norm Internalization

Hypothesis 1. Internalization of aspects of a norm governing usage occurs more rapidly and is easier to achieve than aspects governing development, proliferation, and disarmament. Constraining norms for emerging-technology weapons are more likely to succeed if they address the use rather than the possession of the particular weapon. This may be because such norms allow states to preserve maximum flexibility in pursuing their self-interest because they can still develop and possess the emerging-technology weapons in case a norm fails to constrain the threat. This secondary hypothesis was clearly demonstrated by the relatively early success in establishing constraining norms regarding the use of chemical, biological, and nuclear weapons as well as the permissibility of engaging in strategic bombing compared to constraining norms regarding the possession or proliferation of these various emerging-technology weapons. The main reason for this is that intuitively it is much easier to observe compliance with a usage norm (vs. possession), and thus norm evolution theory's motives of shame, legitimacy, reputation, and esteem are more effective.

Hypothesis 2. Congruent support and involvement from the public and private sectors (particularly industry participants associated with multiuse technology) is key to achieving norm internalization. Internalizing a norm for emerging-technology weapons involving multiuse technology with peaceful applications is made more difficult by the higher number of stakeholders who need to be persuaded by norm entrepreneurs. If these private sector stakeholders are not in synch with their public sector government norm leaders, then internalization of the norm will be more difficult. For example, the lack of support from the biotech and pharmaceutical industries impeded codification of certain aspects of the constraining norm regarding biological weapons, which the presence of support from the chemical industry helped lead to the successful codification of the constraining norm for chemical weapons. Specifically the involvement of the CMA played a major role in the development of the CWC.[24]

Hypothesis 3. Secrecy associated with emerging-technology weapon programs and the possible multiuse nature of their technology will impede norm evolution, especially internalization. If an emerging-technology weapon is based in part on multiuse technology, the involvement of private sector stakeholders will be important to norm internalization. At an even more basic level, however, the multiuse nature of certain weapon technologies impedes the development of a constraining norm because distrustful states in the anarchic international system are reluctant to embrace a normative constraint on a weapon that could easily and covertly be developed by adversaries under peaceful auspices, only to quickly be converted to a belligerent purpose. The multiuse nature of aircraft was a major impediment to success at the Geneva Disarmament Conference of 1932, where delegates struggled to come up with a way to constrain military air forces in the context of the easily convertible commercial aircraft.[25] Additionally, secrecy associated with emerging-technology weapons can make norm internalization difficult due to the difficulty of observing compliance and therefore leveraging actor motives of conformity, legitimacy, reputation, and esteem. After all, if a nation can violate a norm without risk of these pressures, true norm internalization is less likely. Due to the secrecy associated with biological weapon programs (which lack many outwardly observable characteristics), the Soviet Union successfully ramped up their own covert biological weapons program after they signed the BWC and publicly disavowed biological weapons. The Soviet program ultimately became the largest biological weapons program in history, and the Soviet Union managed to keep this violation of the BWC secret for decades.[26]

Hypothesis 4. International pressure for conformity, enabled by real-time media coverage of the weapon's use, will promote internalization. As an element of norm evolution theory's optimal timing factor, real-time media plays a major role in influencing public perceptions. This not only applies to the first two stages of the norm life cycle but also helps create international pressure to conform to internalized norms. For example, in 2013 the media reported on the Syrian government's suspected use of sarin nerve gas in their ongoing civil war.[27] The international community reacted strongly; the United States asserted that this CW attack crossed a "red line" and

Table 9. Hypotheses for norm evolution for emerging-technology weapons

PRIMARY HYPOTHESIS

Direct or indirect alignment of national self-interest with a constraining norm leads to norm emergence, and the extent to which it is aligned with key or powerful states' perception of self-interest will determine how rapidly and effectively the norm emerges.

SECONDARY HYPOTHESES FOR NORM EMERGENCE

1	Coherence and grafting with existing norms will play a key role in the early foundation of the norm for the emerging-technology weapon.
2	Permanently establishing a norm before the weapon exists or is fully capable or widespread will enhance a norm's chance of emergence and eventual cascade.
3	There will be multiple challenges with undemonstrated emerging-technology weapons, specifically: • Differing perspectives as to its future capability, which can impair norm emergence. Threat inflation regarding the possible effects of the weapon (often by the private sector via industry and lobbying groups) can inadvertently strengthen norm emergence for a constraining norm. • Prospect for inadvertent escalation to lack of clarity regarding new technology.
4	Connections with the idea that the weapon can't be defended against will fuel interest in a constraining norm but also limits the effectiveness of reciprocal agreements and can lead to weapon proliferation. Once convention-dependent norms are violated, intense domestic pressure can then build for retaliatory violations of the norm.
5	Initial weapon proliferation and adoption will play a role in norm emergence, specifically: • Unitary dominance of a single actor with the particular weapon type can give that actor significant influence in norm emergence for that weapon type. • Delays in a weapon type's proliferation (often due to technological barriers) can create added time for a constraining norm to emerge. • Varied rates of adoption of the new technology will result in varied interests in the norm and make norm emergence more difficult.

Table 9. (*continued*)

SECONDARY HYPOTHESES FOR NORM CASCADE	
1	Improvements in technology that address previous challenges in adhering to a constraining norm can rapidly lead to a norm cascade.
2	Characterizing the weapon type as unconventional or otherwise granting it a special status can accelerate norm adoption and ultimately achieve a norm cascade.
3	Public demonstrations of the weapon type—enabled by real-time media—can influence public opinion and spread the constraining norm.
4	The international arms control and disarmament bureaucracy and the increasing regulation and legalization of armed conflict provide an increased number of organizational platforms and networks to spread the norm and more rapidly achieve a norm cascade.
SECONDARY HYPOTHESES FOR NORM INTERNALIZATION	
1	Internalization of aspects of a norm governing usage occurs more rapidly and is easier to achieve than aspects governing development, proliferation, and disarmament.
2	Congruent support and involvement from the public and private sectors (particularly industry participants associated with multiuse technology) is key to achieving norm internalization.
3	Secrecy associated with emerging-technology weapon programs and the possible multiuse nature of their technology will impede norm evolution, especially internalization.
4	International pressure for conformity—enabled by real-time media coverage of the weapon's use—will promote internalization.

threatened serious consequences.[28] Ultimately this robust response—including the threat of the use of force to respond to violations of the constraining CW norm—created such pressure to conform to the norm that Syria acceded to the CWC and inspectors began to dismantle Syria's illicit chemical weapons program (although the speed of dismantlement would later become an issue).[29]

Summary

While the general hypotheses of norm evolution theory were for the most part validated in the case studies, the refined norm evolution theory for emerging-technology weapons offers a better and more tailored framework for understanding and predicting norm development in this area. Table 9 summarizes the primary and secondary hypotheses for norm evolution theory for emerging-technology weapons in each stage of the norm life cycle.

6 Predicting Norm Evolution for Cyber Warfare

> The growing use of cyber capabilities to achieve strategic goals is also outpacing the development of a shared understanding of norms of behavior, increasing the chances for miscalculations and misunderstandings that could lead to unintended escalation.
>
> Director of National Intelligence James R. Clapper

Just as constraining norms emerged for emerging-technology airpower weapons early in the twentieth century and for nuclear weapons in the mid-twentieth century, constraining norms are beginning to emerge for emerging-technology cyber weapons. However, the growing international use of cyber weapons is outpacing the development of shared norms of behavior.[1] If constraining cyber norms are not keeping pace, where do they currently stand and how will they evolve going forward? The advent of cyber warfare poses a range of challenges to states. Some of these challenges were also presented by the advent of the other emerging-technology weapons, making norm evolution theory for emerging-technology weapons particularly pertinent for cyber warfare. This chapter offers predictions for the future of constraining norms for cyber warfare by applying this newly refined theory based on an assumption that current trends continue. Various candidate norms for cyber warfare are beginning to emerge through state practice and deliberate norm cultivation efforts as more and more states and other actors begin to grapple with this growing threat through an increasing number of organizational platforms. These early norms are sometimes mutually exclusive and contradictory and include constraints such as limiting targets to military objectives, no first use of cyber weapons, and the application of existing LOAC to cyber warfare. Due to the primary hypothesis as well as the secondary hypotheses for norm evolution for emerging-technology weapons, norm emergence of constraining cyber norms is particularly unlikely should current trends continue. However, if norms do emerge, they should be able to relatively easily achieve a norm cascade, but internalization will again be difficult and is unlikely. (For more general background on cyberspace and cyber warfare, see the appendix.)

Current Status of Cyber Warfare and International Norms

As cyber weapons are emerging-technology weapons and have existed for only a short time and there is relative secrecy surrounding most cyber operations, there is not an extensive record of customary practice of states.[2] Jason Healey breaks the history of cyber conflict into three phases: "realization" in the 1980s, "takeoff" from 1998 to 2003, and "militarization" from 2003 to the present.[3] Two of the main differences in each of Healey's phases are the increasing diffusion of capabilities among nations and improved and formalized organizational approaches to cyber conflict. Focusing on the militarization phase, James Lewis and the Center for Strategic and International Studies (CSIS) maintain a rolling list of "significant cyber incidents" since 2006 and, as of July 2013, identify 153 hostile cyber operations.[4] While Lewis does not separate the operations by CNE and CNA operations, the vast majority (137 of 153, approximately 89 percent) of the incidents appear to be CNE-style operations.[5] Much of the hostile cyber activity to date is not true cyber warfare, but instead is CNE and cyber crime; however, this should not be interpreted as a customary practice against conducting CNA-style cyber attacks.[6] Instead, it is evidence of how early we are in the cyber era (akin to the absence of strategic bombing in the first decade of the nineteenth century) as advanced cyber warfare is only now becoming possible and a robust target set emerging as societies become more immersed and dependent on cyberspace. In the absence of firmly established norms governing cyber warfare, states may also be exhibiting an abundance of caution as they slowly test the limits of what the international community deems acceptable behavior in cyberspace. Of the major CNA-style attacks that have occurred, seven are summarized in table 10, including what they may portend for acceptable norms of behavior in cyberspace as well as the suspected sponsor and the target and effect of the attack. (More detail on these attacks is found in the appendix.)

These seven CNA-style attacks collectively provide some insight into the emergence of international norms through the customary practice of cyber warfare. There are three main takeaways from the attacks. First, the majority (five of seven) of the attacks were aimed at civilian targets, showing that a norm constraining targeting to explicitly military targets or objectives has not yet arisen. Second, to the extent attacks did strike exclusively military targets, they were suspected to have been launched by Western nations (the United States and Israel). This seems to indicate that there may be competing, in some cases more permissive, norms regarding cyber warfare depending on the bloc the nation is associated with—which is consistent

Table 10. Selected CNA-style cyber attacks

ATTACK NAME	DATE	TARGET	EFFECT	SUSPECTED SPONSOR
Trans-Siberian Gas Pipeline	June 1982	Soviet gas pipeline (civilian target)	Massive explosion	United States
Estonia	April–May 2007	Commercial and governmental web services (civilian target)	Major denial of service	Russia
Syrian Air Defense System as part of Operation Orchard	September 2007	Military air defense system (military target)	Degradation of air defense capabilities allowing kinetic strike	Israel
Georgia	July 2008	Commercial and governmental web services (civilian target)	Major denial of service	Russia
Stuxnet	Late 2009–2010, possibly as early as 2007	Iranian centrifuges (military target)	Physical destruction of Iranian centrifuges	United States
Saudi-Aramco	August 2012	State-owned commercial enterprise (civilian target)	Large-scale destruction of data and attempted physical disruption of oil production	Iran
Operation Ababil	September 2012–March 2013	Large financial institutions (civilian target)	Major denial of service	Iran

with the expected competitive environment in the early days of norm emergence outlined by norm evolution theory. Third, experience with cyber warfare is very limited at this point. No known deaths or casualties have yet resulted from cyber attacks, and the physical damage caused, while impacting strategically significant items such as Iranian centrifuges and Soviet gas pipelines, has not been particularly widespread or severe. While the current absence of massively disruptive cyber attacks is likely due to the limited capabilities and transparency or visibility and not a constraining norm, the lack of such attacks may allow space for a constraining norm to emerge.

It is apparent that few, if any, normative constraints governing cyber warfare exist, yet increased attention and discussion, among other things, has helped spur various efforts to reach a consensus on and codify emerging norms for cyber warfare. Norm evolution theory indicates that norm emergence is more likely to occur when key actors are involved, specifically norm entrepreneurs with organizational platforms and key states acting as norm leaders. The two primary intergovernmental bodies and organizational platforms (and subplatforms) currently being used to promote emerging norms for cyber warfare are the UN and NATO. There are some other key multilateral efforts to encourage the development of cyber norms, such as the London Conference on Cyberspace (and subsequent conferences) and academic cyber norm workshops. Efforts in the UN have primarily been led by Russia, and efforts within NATO have been led by the United States.

At the UN the main focus on cyber warfare has occurred in the General Assembly's First Committee (Disarmament and International Security Committee) as well as various subsidiary organs and specialized agencies, particularly the International Telecommunications Union (ITU) of the UN Institute of Disarmament Research (UNIDIR) and the Counter-Terrorism Implementation Task Force (CTITF) working group.[7] There was some movement earlier in the 1990s, but serious focus on cyber warfare began in 1998, when the Russian representatives introduced a resolution in the First Committee titled "Developments in the Field of Information and Telecommunications in the Context of Security" that would begin the process to establish "cyber arms control" similar to other arms control agreements such as the CWC.[8] The Russian proposal is designed to lead to a prohibition of offensive cyber weapons as well as a ban on cyber terrorism, which many interpret as a prohibition on politically destabilizing speech rather than actual cyber attacks.[9] Many interpret Russia's intentions as disingenuous and intended to suppress U.S. cyber superiority as well as internal disagreements and protests leading to "color" revolutions; however, they have been acting as the lead norm entrepreneur and leader for both of these norms.[10] The United States has frequently led opposition to this proposal due to a concern that the proposed treaty would be impractical and unenforceable and the broad scope of "information warfare" would infringe on U.S. favored norms regarding civil liberties and freedom of expression.[11] (This proposal and other efforts within the UN are discussed in more detail in the appendix.) These UN efforts, led primarily by Russia but also now with significant involvement of the United States, Germany, and others, have helped establish various platforms for discussion and fostered the emergence of various norms for cyber activity. Relative to cyber warfare, there have been three key candidate norms: a prohibition on cyber weapons altogether as part of cyber arms control, a prohibition on first-use of cyber weapons, and an obligation to prevent cyber attacks by nonstate actors from originating in their territory. However, these candidate constraining norms have only begun to emerge and gain support. Advocacy on their behalf may be mere diplomatic posturing based on the contemporary practice of cyber warfare, which exhibited no such constraints. Efforts are complicated by the lack of agreement on key terms and concepts, such as whether or not propaganda and information warfare are part of cyber warfare.

NATO has also served as a main intergovernmental body and organizational platform for promoting emerging norms for cyber warfare. Following the major cyber attacks on Estonia (a NATO member) in 2007 and Georgia (an aspiring NATO member) in 2008, NATO began to focus more seriously on the threat of cyber warfare.[12] In 2008 NATO established the NATO Cooperative Cyber Defence Centre of Excellence (NATO CCD COE).[13] The NATO CCD COE is located in Tallinn, Estonia (the epicenter of the 2007 cyber attack) and is sponsored by eleven NATO members. It is focused on enhancing NATO's cyber defense through research, education, and consulting. In 2012 the organization published *National Cyber Security Framework Manual* to help member nations better develop national policies for cyber defense. NATO's commitment to addressing cyber warfare extends beyond this center of excellence. In November 2010 NATO adopted a new strategic concept, which recognized that cyber warfare "can reach a threshold that threatens national and Euro-Atlantic prosperity, security and stability."[14] In general NATO, led by the United States, has approached cyber warfare from a perspective that seeks to apply the existing LOAC to cyber attacks rather than pursue more comprehensive and new restrictions like those proposed by Russia in the UN. NATO's most important activity in this effort was the development of the *Tallinn Manual on the International Law Applicable to Cyber Warfare*.[15] The *Tallinn Manual*, which does not reflect official NATO opinion but rather the personal opinion of the authors (an "international group of experts"), was sponsored by the NATO CCD COE and three organizations acting as observers: NATO, Cyber Command (CYBERCOM), and the International Committee of the Red Cross.[16] Also noteworthy is an independent yet similar effort by Israel, led by Col. Sharon Afek, which reached similar conclusions regarding the LOAC and cyber warfare in early 2014.[17] The *Tallinn Manual* represents not only the consensus view of these NATO-affiliated participants but also the main positions of the U.S. government.[18] This is based on a September 2012 speech by a U.S. State Department legal advisor, Harold Koh, who articulated the U.S. positions on international law and cyberspace, which are consistent with the positions articulated in the *Tallinn Manual*.[19] In addition, the 2011 U.S. "International Strategy for Cyberspace" specified that the "long-standing international norms guiding state behavior—in times of peace and conflict—also apply in cyberspace."[20] Both the Koh speech and the *Tallinn Manual* go further to flesh out the U.S.-NATO position that the international LOAC are adequate and applicable to cyber warfare and reject the Russian position that cyber warfare requires new and distinct international norms and agreements. In addition to the UN and NATO, individual nations have become norm entrepreneurs and norm leaders in organizing ad hoc multilateral forums to discuss norms for cyber warfare, among other things. (These and NATO's efforts in this area are discussed in more detail in the appendix.)

Various candidate international norms for cyber warfare are beginning to emerge as more and more states and other actors grapple with this growing threat. An increasing number of key actors are involved in this norm development process, increasing the likelihood of norm emergence reaching a tipping point, as predicted by norm evolution theory. These actors—motivated by a variety of factors and employing a range of mechanisms—have promoted these norms through a variety of organizational platforms as well as general observance

Table 11. Emerging candidate norms for cyber warfare

NORM	ORGANIZATIONAL PLATFORM(S)	ENTREPRENEUR(S)/ LEADER(S)
Targeting civilian and commercial objectives is acceptable	N/A; State practice, doctrine/strategy	Russia (?), China (?), Iran
Total prohibition on cyber weapons and cyber warfare	UN First Committee	Russia, China
No first use of cyber weapons	UN First Committee, ITU	Russia, China
Responsibility to prevent cyber attacks from a state's territory	ITU; London, Budapest, Seoul conferences	Russia, China, United States, United Kingdom, South Korea, NATO
Cyber CBMs are necessary to prevent misunderstanding	UN GGE, ICT4Peace, World Summit on the Information Society, MIT Cyber Norm Workshops, NATO	Russia, China, United States, United Kingdom, South Korea
Existing LOAC apply to cyber warfare (including limiting targets to narrow military objectives)	UN GGE; NATO; London, Budapest, Seoul conferences; state practice (for limiting targets to military objectives)	United States, United Kingdom, Germany, NATO, Israel

leading to contemporary state practice. Candidate norms and the related organizational platforms and actors are identified in table 11.

Although this growing consensus for various norms is a sign of progress, many challenges remain. Colonel Afek, former Israeli deputy military advocate general, highlighted some of these challenges when he said the international community "faces a complex and challenging period in which we can expect both a cyber arms race with the participation of state and nonstate entities, and a massive battle between East and West over the character of the future legal regime."[21] The remainder of this chapter builds on the current status of norms for cyber warfare by offering various predictions for how constraining norms for cyber warfare will evolve based on norm evolution theory for emerging-technology weapons.

Potential Game-Changing Developments

Before examining the prospects for norm evolution should current trends continue, it worth examining various game-changing developments that could have a disruptive impact and unexpectedly influence norm evolution and thus the predictions that are to follow. The most noteworthy game-changing developments include the introduction of revolutionary technology, a rise of nonstate cyber actors, and the launching of a major strategic cyber attack. These developments are summarized in table 12.

The first potential game-changing development impacting norm evolution would be the sudden arrival of revolutionary cyber technologies. These could be in the form of a breakthrough in quantum computing technology that gives one state a massive advantage in cyberspace, allowing it to attack previously well-defended targets and penetrate advanced cryptology. Gret Tallant, a manager at Lockheed Martin who is working on quantum computing, has said that "computationally, quantum computing is the equivalent of the Wright Brothers at Kitty Hawk; it has the potential to be a turning point in our history."[22] Lockheed's Quantum Computing Center recently upgraded to a 512-qubit machine, and the National Security Agency (NSA) allegedly has a $79.7-million research program focused on these machines, so the technology is not simply speculative science fiction.[23]

Table 12. Game-changing developments that could impact the future of cyber warfare

SCENARIO	DESCRIPTION
Introduction of revolutionary cyber technology	Introduction by one or more actors of revolutionary new cyber technology, which immediately renders current encryption or cyber defenses obsolete.
Rise of nonstate cyber actors	Significant cyber attack capabilities proliferate, and a range of nonstate actors (corporations, hacktivists, terrorists, etc.) increasingly engage in cyber warfare, becoming the dominant cyber actors and blurring attribution and the role of states.
Major strategic cyber attack(s)	A catastrophic attack or series of attacks aimed at civilian targets (power grid, infrastructure, etc.) occurs, leading to physical destruction and loss of life.

However, recent stumbles in keeping pace with Moore's Law may indicate that the pace of technological breakthroughs is slowing.[24] Another potential revolutionary technological breakthrough would be the development of attack code that could spread through sonic transmission (sending messages through acoustic waves between speakers and microphones) to "air-gapped" devices not otherwise accessible.[25] Learning computing technology based on the biological nervous system (referred to as neuromorphic processing) offers a third potential avenue for a major cyber breakthrough.[26] Qualcomm is expected to make the first such processor commercially available in 2014.[27] Quantum computing, sonic malware, and neuromorphic processors all could upend the cyber balance of power and thus affect existing and prospective constraining cyber norms. The abrupt arrival of revolutionary technology would have an uncertain yet potentially profound impact on norm evolution as it would suddenly shift state self-interests in fostering and adhering to constraining cyber norms.

A second potential game-changing development would be a rise in and increasing dominance of hostile nonstate actors in cyberspace. Some individuals, such as the futurist and defense expert Peter Singer, highlight that thus far terrorists and other hostile nonstate actors have not engaged in cyber warfare and that there are some complexities and barriers preventing them from easily conducting sophisticated attacks.[28] However, others, such as Ralph Langner, an expert in industrial control systems, has asserted that vast resources and intelligence capabilities needed to conduct an attack like Stuxnet may not be needed for future attacks on "cyber-physical" systems.[29] A recent report by the U.S. National Intelligence Council (NIC) identified a possible "tectonic shift" whereby individuals and small groups have access to lethal and disruptive technologies, including cyber weapons. These capabilities were previously available only to powerful states. The NIC went on to discuss the plausible scenario and increased risk in the future of cyber mercenaries with critical skills selling their services to terrorists who are interested in causing widespread economic and financial disruptions.[30] In this contingency there is a global diffusion of significant cyber attack capabilities to nonstate actors due to, among other things, the multiuse nature of the technology, increasingly low cost of entry, and growing global reliance on information technology (IT) systems (a growing target set). In addition to the NIC-envisioned cyber terrorists, these actors could include corporations who are "hacking back," politically motivated hacktivists (such as Anonymous), and nationalistic groups with informal state affiliation. Perhaps as a leading indicator of this growing nonstate cyber threat, in late 2013 a Boeing vice president admitted to being "very concerned" about terrorists using malware to conduct a cyber attack against a large airliner due to the numerous IT systems that openly communicate with airport personnel and air traffic control during takeoff and landing.[31] The rise of nonstate actors would potentially impair norm evolution due to extreme proliferation, blurred attribution, and increased prospects of inadvertent escalation. Potential benefits, such as increased pressure for a norm cascade due to widespread public demonstrations of cyber weapons (an expectation developed from the history of norm development for other emerging-technology weapons), would be negated by the limited and challenging application of constraining norms to nonstate actors, who generally are not as likely to be influenced or constrained by norms. While evidence indicates international norms do constrain and influence some armed nonstate actors, these tend to be those who "value their public reputation, moral authority, and source of legitimacy" and essentially perceive themselves as a representative of a distinct population.[32] This applies to only some of the many potential nonstate cyber actors and does not apply to ideologically oriented groups such as Anonymous or al Qaida. These other groups are typically perceived as terrorists or otherwise illegitimate and thus are less inclined to adhere to international norms for legitimacy, reputation, and esteem motives.[33]

A third potential game-changing development would be the occurrence of a major cyber attack or series of attacks on civilian targets, perpetrated by either state or nonstate actors. Such an attack is not deemed implausible by senior leaders. In March 2013 James R. Clapper, the director of national intelligence, testified that in just the next two years there was a very real threat that a major cyber attack against the United States would occur, resulting in "long-term, wide-scale disruption of services, such as a regional power outage."[34] In 2014 Clapper testified that this threat was increasing as "malware and attack tradecraft proliferate."[35] In 2010 the *Economist* envisioned the most extreme of major cyber attacks when it described "the almost instantaneous failure of the systems that keep the modern world turning. As computer networks collapse, factories and chemical plants explode, satellites spin out of control and the financial and power grids fail."[36] The targets of such an attack could include hospitals, Supervisory Control and Data Acquisition (SCADA) control systems for chemical or nuclear plants, water filtration systems, transportation systems such as air traffic management systems and subways, banking and financial systems, and the electrical grid itself.[37] Regarding the electrical grid, the potential consequences could be severe. In 2007 the National Academy of Sciences (NAS) estimated than a major cyber attack on the U.S. electrical grid could lead to "hundreds or even thousands of deaths" due to exposure to extreme temperatures.[38] In May 2013 a report on the grid's vulnerability from Congressmen Edward Markey and Henry Waxman added further credibility to NAS's estimate. The congressional report points out that most utilities are subject to numerous daily cyber attacks, they do not comply with the most robust cyber security standards, and available spare transformers may not be adequate.[39] Doug Myers, chief information officer for Pepco, an electric company in the mid-Atlantic region, predicts that it isn't a question of *if* a cyber attack on the electrical grid happens, but *when*.[40] A study from the U.S. Military Academy's Network Science Center found that cyber attackers can "'cause blackouts by targeting a relative handful of small substations—the often-overlooked and poorly-defended parts of a power grid' . . . leading to a 'chain reaction of power overloading known a cascading failure.'"[41] The government is responding to this looming threat of major cyber attacks on civilian critical infrastructure, for instance hosting a massive public-private exercise called GridEx II in November

2013.[42] In all likelihood such a demonstration of raw cyber power will help spur activity among norm leaders and entrepreneurs, fueling the development of constraining international norms.

Predicting Norm Evolution for Cyber Warfare

Assuming these potential game-changing developments do not occur, it is possible to outline a future scenario for cyber warfare based on current trends.

Likely Future Scenario

This future scenario entails a continuation of the current trends regarding cyber warfare. These trends include cyber conflict becoming more destructive, remaining largely covert with limited public discussion, involving an increasing and continued mix of state and nonstate actors, and more U.S., Russian, Chinese, and Iranian offensive cyber operations (among others). Jason Healey, in *A Fierce Domain: Conflict in Cyberspace*, identifies these developments as the status quo trend based on the past decade of cyber warfare.[43] More destructive and sophisticated cyber weapons are likely, in part due to the success and example provided by Stuxnet and the interest in and proliferation of cyber weapons it has spawned along with the absence of constraining norms on developing such weapons. Stuxnet likely cost in the low double-digit millions of dollars to produce; cyber security expert Eugene Kaspersky has said that given that Stuxnet's code is publicly available, it would be "quite easy to disassemble the ,code to discover how it works, to extract the components and to redesign the same idea in a different way."[44] As a result the cost of cyber weapons is likely decreasing as they (and their associated code) proliferate and are increasingly employed.

Cyber warfare involves a combination of characteristics that make it particularly attractive to states and also fuel its spread. They also help explain why the United States, in spite of its interest in developing constraining cyber norms, has continued to pursue secretive military and intelligence CNA capabilities over the past ten years.[45] These characteristics include the challenges of attribution, the multiuse nature of the associated technologies, target and weapon unpredictability, potential for major collateral damage or unintended consequences, questionable deterrence value, the frequent use of covert programs to develop such weapons, attractiveness to weaker powers and nonstate actors as an asymmetric weapon, and the use as a force multiplier for conventional military operations.[46] Cyber warfare capabilities are leading to a new RMA, wherein cyber capabilities will play an increasingly decisive role in military conflicts and become deeply integrated into states' doctrine and military capabilities. Over thirty countries have taken steps to incorporate cyber warfare capabilities into their military planning and organizations; thus the use of cyber warfare as a "brute force" weapon is likely to increase.[47] This increased interest and incorporation of cyber attack capabilities into the military tool kit could lead to a new RMA wherein offensive cyber operations play an increasingly decisive role in military operations at the tactical, operational, and strategic level. Military planners are actively seeking to incorporate offensive cyber capabilities into existing war plans.[48]

If current trends continue, norm evolution theory as applied to emerging-technology weapons predicts that the constraining norms for cyber warfare will have trouble emerging and may not ever reach a norm cascade. Should the constraining norms manage to successfully emerge, their odds of reaching a tipping point appear better, although internalization is less likely. Of the current candidate cyber norms, the most likely to succeed are those that are more limited, such as those focused on the application of the existing LOAC to cyber warfare or the prohibition on first use of cyber weapons.

Cyber Warfare Norm Emergence

A principal hypothesis of norm evolution theory is that norm emergence is more likely to occur when key actors are involved, specifically norm entrepreneurs with organizational platforms and key states acting as norm leaders. There are a variety of intergovernmental bodies and organizational platforms (and subplatforms) currently being used by a variety of states to promote various emerging norms for cyber warfare. These platforms include the UN, NATO, the London Conference on Cyberspace (and subsequent conferences), and academic venues. Through these platforms a variety of actors, motivated by a number of factors and employing a range of mechanisms, have promoted various candidate cyber norms, ranging from a total prohibition on cyber weapons and warfare to a no first-use policy or the applicability of the existing LOAC to cyber warfare. Norm evolution theory would interpret this as a sign of progress for norm emergence. However, when you dig deeper based on the primary and secondary hypotheses for emerging-technology weapons, the prospects are less hopeful.

Powerful States Unlikely to Support Constraining Norms

The primary hypothesis of norm evolution theory for emerging-technology weapons during norm emergence is that powerful self-interested state actors will play a significant role, and a norm's convergence with perceived state self-interest will be important to achieving norm emergence and a state's acting as a norm leader. In addition to norm entrepreneurs and organizational platforms, successful norm emergence requires states as norm leaders. Since there is generally less exposure and understanding surrounding these cyber weapons as well as different rates of weapon adoption and cyber vulnerability, states will be reluctant to lead on the issue of norms; among other reasons, they may be unable to determine the utility of such weapons relative to their own interests. However, such calculations are essential if important and powerful states are going to become strong norm leaders and help promote the emerging norm. Specific to China, Russia, and the United States—the preeminent cyber actors—an analysis of their respective cyber doctrines and approaches indicates a perspective that each nation has more to gain from engaging in cyber warfare than from significantly restricting it or giving it up entirely.

National investments in cyber warfare capabilities and the development of doctrine and strategies for cyber warfare provide insight into state perceptions of self-interest and the expectations for behavior and emerging constraining norms for cyber warfare. CSIS has identified over thirty countries that are taking steps to incorporate cyber warfare capabilities into their military planning and organizations, and international relations scholar Adam Liff has argued that the use of cyber warfare as a "brute force" weapon is likely to increase in frequency.[49] So where do state cyber warfare programs stand today in China, Russia, and the United States? Unfortunately the same CSIS report indicated that many states keep information about their cyber warfare programs and capabilities secret.[50] However, it is apparent that China, Russia, and the United States are preparing for (and at times engaging in) cyber warfare. There are other important cyber actors, such as North Korea, which are making significant investments in cyber warfare. North Korean leader Kim Jong-Un has been quoted as asserting "that alongside nuclear weapons and missiles, cyber warfare capabilities are 'a magic weapon' that empowers the North Korean army to launch 'ruthless strikes' on the South."[51] Still, the three key states discussed here are the most significant, due to the breadth and sophistication of their capabilities and activities as well as the likelihood that they are serving as the model for many other nations preparing to operate in cyberspace. These states are the key norm leaders that norm evolution theory identifies as important to achieving norm emergence and a norm cascade. Accordingly, reviewing Chinese, Russian, and U.S. interests and approaches to cyber warfare is key to predicting norm evolution.

CHINESE INTEREST IN CYBER WARFARE

China's early activity and interest in cyber warfare indicate that it likely does not consider the emergence of constraining norms as in its self-interest. It has been largely unconstrained by restrictive cyber norms and is preparing to use cyber weapons to cause economic harm, damage critical infrastructure, and influence kinetic armed conflict. As such, it is unlikely to be a vocal norm leader for international cyber norms. China has made expansive efforts in conducting espionage-style cyber operations. In February 2013 the U.S. cyber security firm Mandiant released a study detailing attacks by Chinese military facilities on at least 141 U.S.-affiliated commercial and government targets. Mandiant identified the primarily Chinese actor as "Unit 61398," located within the 2nd Bureau of the People's Liberation Army (PLA) General Staff Department's 3rd department.[52] These attacks prompted the U.S. DOD to classify China as "the world's most active and persistent perpetrators of economic espionage" that is "looking at ways to use cyber for offensive operations."[53] It is this latter point that is of most interest to this book. China is increasingly developing and fielding advanced capabilities in cyberspace. These capabilities are focused not only on collecting sensitive information but also on achieving military effects capable of causing economic harm, damaging critical infrastructure, and influencing the outcome of conventional armed conflicts.[54] China's focus on cyber warfare is a consequence of the potential it holds for traditional Chinese stratagems, especially Sun Tzu's "overcoming the superior with the inferior" (that is, asymmetric warfare) and Chairman Mao Zedong's concept of a "people's war." Chinese military planners view cyber warfare as an important element of the RMA and as a part of their push to perfect the concept of "local war under modern high-technology conditions."[55] The Chinese have been interested in cyber warfare since the advent of the RMA, a military concept focused on warfare fully integrated with information and communications technology (in other words, warfare fully utilizing cyberspace and IT systems).[56] China has been interested in such "informationalized" wars since the early 1990s, when the 1991 Gulf War provided Chinese leaders with a clear example of their importance. Cyber warfare is appealing as a tip-of-the-spear element of China's military forces as it has the potential to degrade a high-tech adversary in a short, high-intensity conflict to the point where the less modern elements of the Chinese forces could play a crucial role. The asymmetric role played by cyber warfare is also obviously appealing to China.[57] The target of this approach is, as former assistant under secretary of defense for policy planning Michael Pillsbury asserts, clearly the United States: "The last 5 years the main Chinese military newspaper *Liberation Army Daily* has published several hundred articles attempting to describe local war doctrine and Chinese military exercises designed to cope with a 'high-tech enemy.' These articles and books leave little doubt that the weapons, equipment, and uniforms that will be possessed by this high-tech enemy will be the forces of the United States or its military allies."[58]

For the Chinese, cyber warfare falls within the paradigm of "unrestricted warfare" discussed by Qiao Liang and Wang Xiangsui, two PLA colonels, in their famous 1999 manifesto *Unrestricted Warfare*. The PLA colonels discuss a form of warfare that "transcends all boundaries and limits" and reflect on the U.S. RMA-driven success in the Gulf War, recognizing the role played by cyberspace. They then take the next step in imagining a future conflict where cyber warfare plays a pivotal role:

> Supposing a war broke out between two developed nations already possessing full information technology. . . . If the attacking side secretly musters large amounts of capital without the enemy nation being aware of this at all and launches a sneak attack against its financial markets, then after causing a financial crisis, buries a computer virus and hacker detachment in the opponent's computer system in advance, while at the same time carrying out a network attack against the enemy so that the civilian electricity network, traffic dispatching network, financial transaction network, telephone communications network, and mass media network are completely paralyzed, this will cause the enemy nation to fall into social panic, street riots, and a political crisis. . . . This admittedly does not attain to the domain spoken of by Sun Tzu, wherein "the other army is subdued without fighting." However, it can be considered to be "subduing the other army through clever operations."[59]

Clearly the authors of *Unrestricted Warfare* apply cyber warfare as a tool to achieve a variation of Sun Tzu's maxim that "subjugating the enemy's army without fighting is the true pinnacle of excellence."[60] Thus Chinese interests in cyber warfare appear to be asymmetric and strategic.

Cyber warfare also appeals to Chinese leaders because it breathes new life into Mao's doctrine of a "people's war."[61] This doctrine arose in the 1970s, when the PLA had over four million members, as an attempt to maximize China's greatest strength—its sheer number of soldiers—by fielding a massive army. It was understood that all elements of the Chinese population would support the force in such a conflict and would be actively engaged either in the militia or, for those territories behind enemy lines, through insurgency tactics. Cyber warfare likewise permits the broader population to participate in future conflicts. Patriotic Chinese citizens can leverage the unique aspects of cyber warfare to participate in an attack anywhere on the globe. This approach is evidenced by the fact that Chinese cyber attacks are often "brute force" attacks that succeed due to their "sheer volume."[62]

Specific Chinese cyber warfare capabilities are difficult to determine due to the lack of transparency regarding their armed forces. What is clear is that China generally does not perceive cyber warfare to be a stand-alone superweapon; rather they view it as a tool to complement and work alongside other military means. This is demonstrated in *Unrestricted Warfare*, which states, "In future warfare, certain advanced weapons may play a leading role. However, as for determining the outcome of war, it is now very difficult for anyone to occupy an unmatched position. It may be leading, but it will not be alone, much less never changing."[63] China has been steadily leveraging its rapidly growing economy to advance its capabilities to act in cyberspace, as noted by Richard Lawless, former deputy undersecretary for defense for Asian and Pacific security affairs. Lawless avers, "Chinese capabilities in this area have evolved from defending networks from attack to offensive operations against adversary networks. . . . [They are] leveraging information technology expertise available in China's booming economy to make significant strides in cyber-warfare."[64] Michael McConnell, former director of national intelligence, testified before Congress that "nations, including Russia and China, have the technical capabilities to target and disrupt."[65] U.S.-China Economic and Security Review Commission reports go further to acknowledge that Chinese cyber warfare capabilities have been incorporated into military doctrine and are capable of achieving large-scale strategic effects: "Chinese military strategists have embraced disruptive warfare techniques, including the use of cyber attacks, and incorporated them in China's military doctrine. Such attacks, if carried out strategically on a large scale, could have catastrophic effects on the target country's critical infrastructure."[66] Chinese military doctrine plans for a combination of cyber and electronic warfare capabilities in the early stages of a conflict.[67] China has actively been developing the capability necessary to make this doctrine a reality. As early as 2003 the PLA organized their first cyber warfare units, likely predecessors to "Unit 61398" (which is in the department responsible for signals intelligence).[68] Since then they have obtained the crucial source codes for Microsoft Office software by leveraging economic access to China to force Microsoft to reveal this sensitive and proprietary information.[69] This information allows the PLA to utilize "zero-day" security flaws in Office applications, which exploit unknown or unpatched software vulnerabilities before the vendor patch is available.[70] This greatly enhances China's ability to plant malicious software designed to collect sensitive information or damage networks and infrastructure. A 2013 report by the cyber security firm FireEye, recognizing these developments, points out that China is the "noisiest" actor in cyberspace and that it had been hacking into critical infrastructure such as dams and 23 gas pipeline companies.[71] In March 2012 the U.S.-China Economic and Security Review Commission released a report titled "Occupying the Information High Ground: Chinese Capabilities for Computer Network Operations and Cyber Espionage," which concluded that "Chinese capabilities in computer network operations have advanced sufficiently to pose genuine risk to U.S. military operations in the event of a conflict."[72] Beyond software-based cyber capabilities and weapons, China has also exploited its status as a producer of cheap IT hardware components, which can be used for hardware-based CNE and even CNA.[73] China is benefiting from massive CNE operations in cyberspace and, perhaps as a result, appears lukewarm at best to constraining norms for the related threat of cyber warfare, as indicated by its limited participation in actively supporting UN efforts led by Russia.

RUSSIAN INTEREST IN CYBER WARFARE

Like China, Russia's early cyber warfare activity—especially the attacks on Estonia and Georgia—indicates that it is largely unconstrained by restrictive cyber norms and is preparing to use cyber weapons in a wide range of conflicts and against a variety of targets. It likely does not consider the emergence of constraining norms as in its self-interest. As such, one would think it unlikely to be a vocal norm leader for international cyber norms (although it was a public advocate of the BWC and CWC while simultaneously seeking to violate the norms those treaties codified). However, Russia has been a leading proponent of a total ban on cyber weapons, just as the Soviet Union attempted to demonize U.S. possession of nuclear weapons while simultaneously pursuing such weapons themselves. This helps illustrate how powerful states acting on their own self-interest can inadvertently act as norm leaders in spite of flouting the candidate norm themselves. However, Russia's confusing support for fully constraining norms for cyber warfare (based on its behavior in the UN and proposal for an "International Code of Conduct for Information Security") may be based on its broader definition of cyber warfare and its interest in using a constraining norm to prevent what it perceives as "propaganda" inside Russia and in its near abroad.[74] Its position may also be disingenuous, as it was with the BWC and the CWC. To achieve any real convergence of the main cyber actors the authoritarian interest in cyber norms to constrain free speech will have to be addressed, which could deflate Russian support.

The Russian Federation was the suspected cyber aggressor in the attacks against Estonia and Georgia, so its general interest in cyber warfare is widely known. However, outside of these and a few other attacks, little is known of Russia's cyber capabilities; some believe Russia is a "little bit too quiet" and that the lack of notoriety is indicative of a high level of sophistication that enables Russian hackers to evade detection.[75] That said, there are some indicators of Russian intent and capabilities. Russia has updated its military doctrine in 2010 to, among other things, emphasize a greater role for cyber and information warfare.[76] Russian doctrine now states that future conflict will entail the "early implementation of measures of information warfare to achieve political objectives without the use of military force, and in the future to generate a favorable reaction of the international community to use military force."[77] Russia previously defined information warfare—their term for cyber warfare—as having four components: "the destruction of command and control centers and electromagnetic attack on information and telecommunications systems; the acquisition of intelligence; disruption of computer systems; and disinformation."[78] As noted previously, senior U.S. intelligence officials claim that Russia has the technical capability to "target and disrupt" in cyberspace, and, like China, has established cyber warfare–focused units within its military.[79]

In 2012 President Vladimir Putin recognized the potential of cyber warfare when he said "space-based systems and IT tools, especially in cyberspace, will play a great, if not decisive role in armed conflicts. . . . All this will provide fundamentally new instruments for achieving political and strategic goals in addition to nuclear weapons."[80] While Russia appears to have some of the most sophisticated (and difficult to detect) cyber capabilities, ironically its efforts have largely been focused inward, on former Soviet states, most recently with apparent cyber attacks on Ukraine and possibly NATO following the crisis over Crimea in March 2014.[81] In June 2013 Russia and the United States signed a "cyber pact" establishing a communication hotline between the two countries in the event of a crisis in cyber space.[82] This is not particularly surprising or indicative of a major breakthrough on cyber norms given the two nations' experience dealing with uncertainty as the world's major nuclear superpowers.

U.S. INTEREST IN CYBER WARFARE

While China is perhaps the noisiest and Russia the most secretive when it comes to cyber warfare, the United States is the most sophisticated. The United States is in the process of dramatically expanding its military organization committed to engaging in cyber warfare and regularly engages in "offensive cyber operations."[83] However, unlike Russian attacks and Chinese planning, it appears to exercise restraint and avoid targeting nonmilitary targets. This seems to indicate that the United States is acting as a norm leader for at least a certain category of constraining cyber norms, although its general "militarization" of cyberspace may be negating the norm-promoting effects of this restraint. The Stuxnet attack is believed to be of U.S. origin and represents the most highly engineered CNA-style attack to date.[84] The United States has focused on cyber threats since the

1990s, and while the Department of Homeland Security and Federal Bureau of Investigation (among other agencies) have defensive roles, the key offensive cyber warfare roles are filled by the military and CIA.[85] The key organization involved in CNA-style cyber operations is CYBERCOM, a military subcommand under U.S. Strategic Command. CYBERCOM recently outlined an ambitious plan to field over one hundred cyber teams by late 2015, including Cyber Combat Mission Forces, responsible for working with geographic combatant commands to conduct offensive cyber attacks.[86] The fiscal year 2014 funding for DOD more than doubled the year-over-year funding for CYBERCOM, from $191 million to $447 million.[87] Purported evidence leaked in August 2013 that the U.S. conducted 231 offensive cyber operations in 2011, further demonstrating advanced offensive cyber capability.[88]

While the United States has recently developed classified rules of engagement for cyber warfare, it has articulated few, if any, limits on its use of force in cyberspace or response to hostile cyber attacks. The May 2011 "International Strategy for Cyberspace" states that the United States "reserves the right to use all necessary means" to defend itself and its allies and partners, but that it will "exhaust all options before [the use of] military force."[89] Additionally, former deputy secretary of defense William Lynn clearly asserted that "the United State reserves the right, under the law of armed conflict, to respond to serious cyber attacks with an appropriate, proportional, and justified military response."[90] In July 2011—around the same time as Lynn's statement—DOD unveiled the unclassified version of its strategy for operating in cyberspace.[91] This unclassified document lacked a lot of the detail presumably found in the classified version; nevertheless the media has provided additional context based on various sources. One key concept disclosed is the idea of "equivalence" to decide when a cyber attack would trigger a conventional response.[92] Officials have indicated that a cyber attack resulting in death, damage, or a high level of disruption comparable to what could be caused by a kinetic military attack would be grounds for a conventional response. Perhaps more provocatively, in 2013 the DOD Defense Science Board implied the United States would use noncyber weapons, including even nuclear weapons, to retaliate after a major cyber attack.[93] These statements, along with the suspected U.S. cyber attacks, have led many to question whether U.S. militarization of cyberspace is counterproductive to stability and norm emergence.[94] However, DOD recently developed new classified standing rules of engagement (SROE) for cyber warfare, which allegedly permits cyber attacks without National Security Council approval (which previously was required).[95] There is concern that this shift toward more regular and normalized cyber warfare, enabled by the new SROE, will spread to other militaries. Ultimately U.S. behavior and interest in cyber warfare indicate that it does not consider the emergence of robust constraining norms in its self-interest.

The Snowden Leaks

The Snowden leaks may have introduced more distrust than already existed among adversaries and allies to complicate and hamper a convergence of states' self-interest. On June 5, 2013, the *Guardian*, a British newspaper, began reporting on classified documents provided by Edward Snowden, a former employee of the CIA and contractor at NSA, including documents outlining U.S. offensive cyber attacks conducted against other nations.[96] Suddenly the spotlight was on U.S. cyber activity and the breadth and nature of its secret offensive actions in cyberspace. These purported revelations of classified material, such as the alleged evidence that the United States conducted 231 offensive cyber operations in 2011, are having an impact on international cooperation.[97] The purported revelations regarding the extent of the NSA's cyber intelligence collection efforts have led some U.S. allies, such as Germany and Finland, to construct their own independent IT infrastructure, including fiber optic cables.[98] France has launched its own data countersurveillance efforts, and Brazil's president canceled a state visit to the United States and decried the NSA activities as "an assault on national sovereignty."[99] This led David DeWalt, chairman of FireEye, to predict that there will be increasing "cyber balkanization" with more cyber nationalism and less international cooperation.[100] The current Snowden leaks alone will likely have an impact on the evolution of constraining cyber warfare norms, and more leaks are likely coming. Glenn Greenwald, the journalist who helped Snowden leak his allegedly classified information, said in October 2013 that the full breadth of information Snowden provided is "so complex and so deep and so shocking, that I think the most shocking and significant stories are the ones we are still working on, and have yet to publish."[101] The current and future leaks could fracture state interests and increase national secrecy of cyber weapon programs and distrust of U.S. intentions and those of other powerful cyber actors. This type of effect was evidenced by a Russian government source claiming in late 2013 that "Washington has lost the moral authority" in cyberspace and that support for the Russian UN First Committee cyber resolution was growing and the Group of Government Experts becoming more Russian-friendly. Former Canadian ambassador Paul Meyer noted that the leaks have had a direct impact on UN cyber norm activity and have led some of the more offended nations, such as Germany and Brazil, to call for action to prevent "cyberspace from being used as a weapon of war."[102] The leaks had an impact in China as well, where hostile operations in cyberspace were on the decline following the public shaming resulting from the Mandiant report; after the Snowden revelations and U.S. embarrassment, the Chinese have stepped up their hostile activity to unprecedented levels.[103]

Looking forward, an NIC report on global trends in the coming decades articulated the possibility that "fear of the growth of an Orwellian surveillance state may lead citizens particularly in the developed world to pressure their governments to restrict or dismantle" IT systems, which would include potential cyber weapons.[104] The bottom line is that powerful support from self-interested actors has not converged on a comprehensive constraining norm for cyber warfare, and recent developments may make such a convergence less likely. Perhaps the best hope is a more limited cyber norm applying the existing LOAC to cyber warfare.

Coherence with Existing Dominant Norms Unlikely

Of the five secondary hypotheses for norm emergence developed in chapter 5, the majority are not favorable when applied to cyber warfare should current trends continue. First, cyber norms will have difficulty achieving coherence with and grafting onto existing norms. Unfortunately, the success of a norm candidate for emerging-technology weapons will depend in large part on the ability to achieve coherence by connecting the new weapon type to an existing category and thus beginning the process of grafting the new norm onto existing norms. While cyber weapons and cyber warfare have some commonalities with certain weapons, particularly unconventional and emerging-technology weapons, overall they are truly unique. In fact they are so unique as to operate in their own new, man-made domain and outside of the normal domains of land, sea, air, and space. As such, cyber norms lack obvious coherence with many prominent norms;

thus it is difficult for norm entrepreneurs to actively seek to graft the candidate norms to existing norms. Perhaps the best option for grafting cyber norms is the humanitarian norm underlying the existing LOAC, particularly the norm regarding the protection of civilians and minimization of collateral damage.[105] This is precisely what NATO's *Tallinn Manual* attempts to achieve by arguing that the LOAC apply to cyber warfare.[106] However, the lack of agreement on key terms and confusion over the spectrum of hostile cyber operations make coherence and grafting complex and difficult.[107]

Too Late for Preemptive Establishment of Norms

Another secondary hypothesis for norm emergence for emerging-technology weapons is that it will be more successful if the candidate norm can be permanently and preemptively established before the weapon exists or is fully capable or widespread. With cyber warfare, the train has already left the station: Lewis and CSIS identified sixteen significant CNA-style cyber attacks between 2006 and 2013.[108] These included major attacks in the former Soviet states of Estonia and Georgia, and in Iran and Saudi Arabia. While no one has yet been killed by a cyber attack, the opportunity for permanent preemptive establishment of a norm has long since passed.

The Undemonstrated Nature of Cyber Warfare

Norm evolution theory for emerging-technology weapons indicates that with undemonstrated emerging-technology weapons, there will be challenges arising from both differing perspectives of their future capability as well as the prospect for inadvertent escalation. While it is true that cyber warfare has been demonstrated to some degree (Stuxnet, etc.), the hidden and secretive nature of cyberspace makes the actors and their intent and objectives unclear and thus limit the true demonstrative value of recent cyber attacks. This has the effect of cyber weapons being largely "undemonstrated" and leaves ample room for competing theories and arguments on their effectiveness and strategic impact. Some analysts, policymakers, and academics argue that cyber warfare poses a major threat and warn of a "cyber Pearl Harbor" or "cyber 9/11," when critical infrastructure is attacked. Advocates of the impact and severity of the threat of cyber warfare have included leading decision makers, such as U.S. Secretary of Defense Leon Panetta, who warned in October 2012 of an attack in which "an aggressor nation or extremist group could use these kinds of cyber tools to gain control of critical switches. . . . They could derail passenger trains, or even more dangerous, derail passenger trains loaded with lethal chemicals. They could contaminate the water supply in major cities, or shut down the power grid across large parts of the country."[109]

On the other side of the spectrum, some have argued that statements such as Panetta's are pure hyperbole, that in fact cyber warfare may not even constitute warfare as properly defined. The German academic Thomas Rid is the leading advocate of this argument; he makes the case in his popular book *Cyber War Will Not Take Place*, and in a December 2013 article in *Foreign Affairs*, where he asserts that a cyber attack is not a major threat but will in fact "diminish rather than accentuate political violence" by offering states and other actors a new mechanism to engage in aggression below the threshold of war.[110] Others, such as international relations scholar Erik Gartzke, share Rid's view and argue that cyber warfare is "unlikely to prove as pivotal in world affairs . . . as many observers seem to believe."[111] Norm evolution theory as applied to emerging-technology weapons indicates that these vastly different perceptions of the impact and role of cyber warfare in international relations and conflict will impair norm emergence.

The lack of understanding regarding cyber weapons resulting from the lack of demonstration and inherent secrecy of such weapons also increases the prospects for inadvertent escalation and misinterpretation. Years after the Stuxnet attack, analysts are still trying to determine the precise intent and plan of the attackers.[112] Additionally cyber security is a huge and booming business for IT-security firms; the industry analyst Deltek reported that the U.S. government IT-security market will increase from $8.6 billion in 2010 to $13.3 billion in 2015 (a compound annual growth rate of 9.1 percent).[113] Bruce Schneier, an IT-security expert, has alleged that firms benefiting from cyber growth have, along with their government customers, artificially hyped the cyber threat, using the lack of standard terms or understanding of cyber warfare to conflate a wide range of cyber threats (CNE, CNA, cyber crime, etc.).[114] Some critics have gone so far as to refer to this dynamic as "cyber doom" rhetoric or a "cyber security-industrial complex," similar to the oft-derided "defense-industrial complex."[115] The norm emergence experience for other emerging-technology weapons indicates that this threat inflation to fuel profits may also help fuel the development of constraining norms by motivating norm entrepreneurs and actors. For cyber warfare, this certainly appears to be the case and will likely continue.

The Perception That Cyber Weapons Cannot be Defended Against

A fourth hypothesis is that connections with the idea that the weapon can't be defended against will fuel interest in a constraining norm but also limits the effectiveness of reciprocal agreements and can lead to weapon proliferation. As a result, once convention-dependent norms are violated, intense domestic pressure can then build for retaliatory violations of the norm. Defenses against cyber weapons are largely viewed as inadequate. The DOD's Defense Science Board reported in January 2013 that the United States "cannot be confident" that critical IT systems can be defended from a well-resourced cyber adversary.[116] The nature of cyberspace, with intense secrecy and "zero-day" vulnerabilities, makes defense particularly difficult and fuels interest in other strategies to manage the threat, including constraining international norms. This explains the broad range of actors and organizational platforms involved in early norm promotion and is a positive factor for the successful emergence of norms for cyber warfare. However, the experience of norms for emerging-technology weapons with similar perceptions regarding the weakness of defenses also indicates that while this may fuel interest in cultivating norms, such norms will be fragile and largely apply to use and not proliferation as actors will continue to develop and pursue the weapons as they believe they cannot rely on defenses and seek deterrence-in-kind capabilities. Further, if the early norm is violated, given the inability to defend against continued violations there may be domestic pressure to respond in kind, leading to a rapid erosion of the norm. Should early cyber norms be violated, such domestic pressure for an in-kind response could build. In fact the Iranian attack on Saudi Aramco in August 2012 is largely viewed as one of Iran's responses to Stuxnet.[117] The challenge of attribution in cyberspace may accentuate this dynamic by making retaliatory responses even easier than with prior emerging-technology weapons.

The Proliferation and Adoption of Cyber Weapons

The final secondary hypothesis is that weapon proliferation and adoption will play a significant role in norm emergence as it will influence the primary hypothesis and state interest in constraining norms. Cyber warfare does not admit the unitary dominance of a single actor as there was with the U.S. nuclear monopoly early in the age of nuclear—giving the United States significant influence on norm emergence regarding nuclear restraint. Given the ongoing proliferation of cyber weapons, the multiuse nature of the technology, and the relatively low cost of entry, delays in the proliferation cyber weapon (creating added time for a constraining norm to emerge) are unlikely. However, there will likely be varied rates of adoption of cyber weapons, with some nations (the United States, China, Russia, and Israel) possessing the most sophisticated cyber warheads.[118] Experience with norm development for emerging-technology weapons indicates that states with powerful cyber weapons are more likely to resist the emergence of any constraining norms. This is especially true with strong bureaucratic actors, such as NSA in the United States and the Federal Agency of Government Communications and Information in Russia, potentially advocating for permissive norms. As discussed earlier, while the Russians have been major advocates in the UN for a total prohibition on cyber weapons, their interest may be driven by a perception that the United States is the dominant cyber power, or, perhaps more cynically, it could be akin to the Soviet Union's disingenuous early promotion of the constraining BW and nuclear norms while simultaneously pursuing biological and nuclear weapons. Regardless, the varied rates of adoption and development of cyber capabilities indicate that there will be divergent perspectives on constraining norms, making consensus difficult. This helps explain why, despite the many actors and organizational platforms involved in developing candidate norms for cyber warfare, they have not been successful in achieving any broad consensus beyond the budding consensus regarding the application of the LOAC.

There are also different levels of cyber vulnerability: the more connected and reliant a nation is on IT systems, the more vulnerable it is to cyber attack. Former White House cyber advisor Melissa Hathaway has articulated this vulnerability on a "cyber readiness index," which incorporates national cyber vulnerability and cyber security to come up with a net vulnerability.[119] Vulnerability (perceived or actual) is likely to fuel interest in constraining norms; Hathaway's initial cyber readiness index, which assesses thirty-five countries, indicates that vulnerability is widespread. Again, this helps explain the broad interest in cyber norms and could prove helpful in motivating a consensus on norms down the road, although convergence on a consensus is more complicated.

If current trends continue, the primary and secondary hypotheses of norm evolution theory for emerging technology weapons predicts that the emergence and early development of constraining norms will be challenged and may not occur at all. Key states—especially China, Russia, and the United States—are unlikely to perceive the emergence of robust constraining norms in their self-interest. Further, limited options for coherence and grafting, inability to preemptively establish a prohibition, undemonstrated capabilities, the proliferation and adoption of cyber weapons, and the lack of powerful self-interested state actors converging on a candidate norm present serious hurdles for norm emergence. However, the connection with the idea that cyber weapons cannot be adequately defended against as well as industry and government hyping of the threat have spurred significant general interest in constraining norms for cyber warfare, leading to the many actors and organizational platforms identified in chapter 5. To move past this point and successfully achieve norm emergence, a consensus on cyber norms will need to build, but such a consensus seems unlikely at this point or in the near future. The best and most reasonable candidate is the more limited and modest norm applying the existing LOAC to cyber warfare.

Cyber Warfare Norm Cascade

While the odds of constraining cyber warfare norms achieving norm emergence are not good, should norm emergence occur, it is important to examine what norm evolution theory for emerging-technology weapons predicts regarding achieving a norm cascade. General norm evolution theory hypothesizes that a norm is more likely to reach a tipping point and achieve a norm cascade when key actors such as states and international organizations are involved and motivated by legitimacy, reputation, and esteem (in addition to their obvious motivation of perceived self-interest). Norm evolution theory further suggests that institutionalization, socialization, demonstration, and optimal timing are important to a norm achieving a cascade, as is whether or not a norm is clear and specific and makes universal claims. For cyber warfare norms, there are many actors; however, they are not yet motivated by legitimacy, reputation, and esteem. States are still largely motivated by their own self-interest, and the general altruism and empathy toward mitigating the potential cyber threat does not balance out their perceived self-interest in cyber warfare. Norm entrepreneurs are attempting to use the mechanisms of institutionalization and socialization to promote their respective candidate norms in venues such as the UN and NATO, but such efforts have thus far failed to achieve a broad consensus beyond a general agreement that norms are needed. Finally, norms for cyber warfare at this point are anything but clear and specific given the lack of clarity and agreement on key terms and concepts (such as how to define cyber warfare and what constitutes the use of force in cyberspace). These dynamics lead to pessimistic conclusions about the chances of constraining cyber norms achieving a norm cascade. However, the specific secondary hypotheses for emerging-technology weapons point to better odds for achieving a norm cascade should the candidate norm reach the second stage of the norm life cycle. Of the four hypotheses developed in chapter 5, the majority favor a norm cascade for constraining cyber norms.

Improvements in Technology

Improvements in technology can address previous challenges in adhering to a constraining norm and can rapidly lead to a norm cascade. For cyber warfare, a major impediment to the development of constraining norms is the inherent anonymity built into cyberspace and the related challenge of attribution of cyber attacks. This impedes norm evolution in a number of ways, such as limiting actor motives since actors can reasonably believe they will not be conclusively identified as the responsible party for violating the norm. However, improvements in technology may be able to significantly mitigate the current attribution challenge and thus remove this impediment and influence state perception that norms are in their interest (the primary hypothesis). In late 2012 U.S. Secretary of Defense Leon Panetta said that U.S. cyber attribution skills had improved to the point where they could locate and hold accountable those who harm the United States in cyberspace.[120] Robert Knake at the Council on Foreign Relations said that attribution is increasingly less challenging, even without resorting to labeling each data packet with unique identifiers (a political nonstarter).[121] Should technology reach a point where timely and accurate attribution of malicious cyber actors—particularly those engaging in large-scale cyber attacks—is possible, this would help a constraining cyber norm more rapidly reach a tipping point.

Characterizing Cyber Weapons as Unconventional

Characterizing a weapon type as unconventional or otherwise granting it a special status can accelerate norm adoption and ultimately achievement of a norm cascade. Given the unique combination of special characteristics associated with cyber warfare and the fact that it occurs in a new and man-made domain, it is by default unconventional. Among those who have suggested that cyber weapons be considered another category of unconventional WMD is Geoffrey Ingersoll, who makes the case in his article "The Next Weapon of Mass Destruction Will Probably Be a Thumbdrive."[122] This association of cyber warfare with WMD is pervasive enough to have led the *Bulletin of the Atomic Scientists* to publish an article titled "The Misunderstood Acronym: Why Cyber Weapons Aren't WMD."[123] This ongoing characterization of cyber weapons as different, special, or otherwise unconventional will help isolate cyber weapons and therefore enable the development of special, constraining norms of behavior.

Likely Future Public Demonstrations of Limited Value

With the increasing tempo and frequency of cyber attacks, the possibility of a truly public demonstration of cyber warfare is possible. However, the inherent secrecy and anonymity of cyberspace and cyber conflict make such a demonstration more difficult to achieve. Demonstrations of cyber weapons and cyber power may be more muted than the media-reported nuclear tests on Bikini Atoll or the Syrian use of CW. Some believe the United States has been deliberately leaking information about its alleged cyber attacks, especially Stuxnet, is part of a deliberate effort to demonstrate cyber capabilities, but this subtle demonstration is likely oriented toward influencing state actors and will not be as effective in generating broad public awareness and support for constraining cyber norms.[124]

Increased Legalization of Armed Conflict and International Bureaucracies

War and wartime conduct are increasingly governed by international law and agreements, and a multitude of international organizations exist solely to address security issues and facilitate international discussions. Aside from the UN General Assembly's First Committee, the ITU, UNIDIR, the CTITF working group, NATO CCD COE, and the Red Cross, there are regional multilateral organizations such as the Organization for Security and Co-operation in Europe and the Association of Southeast Asian Nations Regional Forum.[125] These many venues are positive variables that make achieving a norm cascade more likely.

Together, these factors indicate that should constraining cyber norms manage to achieve norm emergence and approach a norm cascade, a tipping point may actually be more likely.

Cyber Warfare Norm Internalization

Norm internalization usually occurs many years or decades after initial norm emergence. For the norm internalization stage in the norm life cycle, cyber weapons (as an emerging-technology weapon) may no longer be considered "emerging technology" and thus will likely be subject to the same prospects as any other norm. The exceptions to this will be contingent on norms for an emerging-technology weapon progressing rapidly through the preceding two stages such that the weapon is still truly novel and different in distribution or conception from other weapon types. Regardless, should constraining cyber warfare norms successfully navigate the long odds of norm emergence and then fare better in achieving a norm cascade, what does norm evolution theory for emerging-technology weapons predicts regarding norm internalization? General norm evolution theory offers various hypotheses regarding what increases a norm's chances of becoming deeply enshrined and achieving taken-for-granted status. First, it predicts that internalization is more likely if the key actors from bureaucracies, the legal community, and important professions are involved. It also hypothesizes that norm internalization is more likely when actors are motivated by conformity and domestic turmoil and when the environment provides expansive professional networks. Cyber warfare norms in the process of internalization would benefit from the presence of an expansive international arms control and disarmament bureaucracy and the increasing legalization of armed conflict. States are likely to be motivated by pressure to conform with the norm once it reaches a tipping point, but domestic turmoil may actually weaken internalization of the cyber norm as states appear to be increasingly engaging in cyber warfare during internal conflict with a likely spillover into their international behavior. For example, the Syrian government is increasingly engaging in CNA- and CNE-style cyber operations against internal rebel groups.[126] Additionally, the involvement of key professions through various professional networks will be difficult. These general hypotheses, when applied to cyber norms, seem to indicate about even odds of internalization. When considered along with the specific hypotheses for emerging-technology weapons, a different conclusion is reached, and the prospects for internalization of constraining cyber warfare norms are grim. Of the four secondary hypotheses, all but one point to challenges for internalization of comprehensive constraining cyber norms.

Norms Governing Usage

The experience of norm internalization for emerging-technology weapons indicates that internalization of aspects of a norm governing usage occurs more rapidly and is easier to achieve than aspects governing development, proliferation, and disarmament. This would indicate that internalization of any constraining cyber norms would likely be limited to the use of such weapons and not their possession or proliferation. The candidate norms for the application of the existing LOAC to cyber warfare or the prohibition on first use of cyber weapons are therefore more likely to succeed in achieving internalization should they make it through the earlier two phases of the norm life cycle. The Russian and Chinese proposal of a total prohibition of cyber weapons is much less likely to fare well.

Unlikely Congruent Support from Public and Private Sectors

The historic experience with internalization of norms for other emerging-technology weapons indicates that congruent support and involvement from the public and private sectors (particularly industry participants associated with multiuse technology) is important to achieving norm internalization. For cyber warfare, the private sector entails the broad category of corporations involved in IT-systems and Internet services. These include the main Internet service providers responsible for providing millions of individuals and entities with Internet connectivity as well as online service companies such as Microsoft, Yahoo, Google, Facebook, AOL, Skype, and Apple. Cyberspace is literally a "private" domain in that it is largely owned and maintained by these private entities. Cyber attacks and cyber warfare occur in this space, and internalization of international norms to constrain cyber behavior will require the participation of and convergence of not only the various government perspectives but also these private entities. Unfortunately such partnership is made less likely after the leaking of allegedly classified information by Edward Snowden.[127] This is because information in the leaks, especially the discussion of the purported PRISM program in which the U.S. government allegedly partnered with private Internet companies to collect information, has led to public outcry based on privacy and civil liberty concerns.[128] Further, the leaks and underlying participation of private corporations in government intelligence collection threaten to fracture industry globally, with different countries adopting different standards. Yahoo CEO Marissa Mayer warned President Obama about this potential backlash in December 2013.[129] The Snowden leaks put these companies under tremendous pressure for their cooperation with government intelligence-gathering efforts, making them less likely to be willing to cooperate with the government on future cyber initiatives, such as internalization of constraining norms, in spite of the fact that they would be among the biggest beneficiaries of such norms. Such internalization would likely entail additional monitoring of private IT systems and would fall under the shadow and suspicion of PRISM. Internalization of constraining cyber norms is therefore made more difficult by the likely lack of congruent support and involvement from the public and private sectors.

Secrecy and Multiuse

The final two secondary hypotheses are that the secrecy associated with emerging-technology weapon programs and the possible multiuse nature of their technology will impede norm evolution, especially internalization, and that international pressure for conformity—enabled by real-time media coverage of the weapon's use—will promote internalization. Cyber warfare's abundance of secrecy and multiuse technology suggest that internalization of the candidate cyber warfare norms will be less likely. The candidate norms requiring more openness and confidence-building measures, promoted in a variety of international venues, may help ameliorate the secrecy challenge by the time the comprehensive constraining norm nears internalization, but that is yet to be seen. The expectation that international pressure for conformity will promote internalization is also eroded by the secrecy and anonymity associated with cyber warfare; thus such a pressure is unlikely to occur.

Collectively, these hypotheses indicate that should constraining cyber norms manage to achieve norm emergence and a norm cascade, internalization will be less likely. This is largely due to the pronounced secrecy and multiuse nature of cyber technologies, which in addition to posing their own barriers to internalization also help blunt international pressure for conformity and private sector support. Norm internalization is likely to be most successful for any norm governing usage rather than development, proliferation, and disarmament. Thus the current candidate norm for the application of the existing LOAC to cyber warfare or is the most likely to be internalized.

Summary

Cyber warfare is still in its infancy, and there are multiple possibilities and contingencies for how this new mode of warfare plays out over the coming decades. However, reasonable conclusions can be drawn regarding the most likely future scenario for cyber warfare. These conclusions—based on norm evolution theory for emerging-technology weapons—indicate that there are many hurdles facing the development of constraining norms for cyber warfare. This chapter identified multiple candidate norms for cyber warfare that are beginning to emerge. However, norm evolution theory for emerging-technology weapons predicts that if current trends continue, constraining norms for cyber warfare will have trouble emerging and may not ever reach a norm cascade. This is principally due to the fact that powerful state actors are unlikely to perceive a convergence between a robust constraining norm and their self-interest. Should constraining norms manage to successfully emerge, their odds of reaching a tipping point are better, although internalization is less likely. Of the current candidate cyber norms identified in this chapter, the most likely to succeed are those that are more limited, such as those focused on the application of the existing LOAC to cyber warfare or the prohibition on first use of cyber weapons. Table 13 summarizes the implications (positive or negative) of norm evolution theory for emerging-technology weapons on the likely future of constraining norms for cyber warfare.

While norm evolution theory for emerging-technology weapons predicts grim prospects for the evolution of constraining cyber norms, unfortunately the threat of cyber warfare is not diminishing and the need for such constraining international norms remains. James Comey, the director of the Federal Bureau of Investigation, testified before the Senate Homeland Security Committee in November 2013 that the risk of cyber attacks is likely to exceed the danger posed by terrorist networks as the top national security threat to the United States: "We have connected all of our lives—personal, professional and national—to the Internet. . . . That's where the bad guys will go because that's where our lives are, our money, our secrets."[130] Former secretary of defense Leon Panetta, speaking at a Symantec Government

Table 13. Norm evolution theory for emerging-technology weapons' implications for norms for cyber warfare

PRIMARY HYPOTHESIS		IMPLICATIONS FOR CYBER NORMS
Direct or indirect alignment of national self-interest with a constraining norm leads to norm emergence, and the extent to which it is aligned with key or powerful states perception of self-interest will determine how rapidly and effectively the norm emerges.		Negative
SECONDARY HYPOTHESES FOR NORM EMERGENCE		IMPLICATIONS FOR CYBER NORMS
1	Coherence and grafting with existing norms.	Negative
2	Permanently establishing a norm before the weapon exists or is fully capable or widespread.	Negative
3	Undemonstrated emerging-technology weapons.	Negative
4	Connections with the idea that the weapon can't be defended against.	Positive
5	Initial weapon proliferation/adoption.	Negative
SECONDARY HYPOTHESES FOR NORM CASCADE		IMPLICATIONS FOR CYBER NORMS
1	Improvements in technology.	Positive
2	Characterizing the weapon type as unconventional or otherwise granting it a special status.	Positive
3	Public demonstrations of the weapon type, enabled by real-time media.	Negative
4	The international arms control and disarmament bureaucracy and the increasing regulation and legalization of armed conflict.	Positive
SECONDARY HYPOTHESES FOR NORM INTERNALIZATION		IMPLICATIONS FOR CYBER NORMS
1	Internalization of aspects of a norm governing usage rather than aspects governing development, proliferation, and disarmament.	Positive
2	Congruent support and involvement from the public and private sectors.	Negative
3	Secrecy and the multiuse nature of the technology.	Negative
4	International pressure for conformity, enabled by real-time media coverage of the weapon's use.	Negative

Symposium in March 2014, said that cyber warfare could "devastate our critical infrastructure and paralyze our nation" and is "the most serious threat in the 21st century."[131] What, if anything can be done to foster and cultivate viable constraining norms for cyber warfare in light of these relatively grim prospects?

Conclusions and Recommendations

In January 2014 a survey of senior U.S. national security leaders identified cyber warfare as the most serious threat facing the United States, and a Pew Research Center poll found that Americans considered cyber attacks to be one of the top threats.[1] Cyber warfare poses a real and growing threat, a threat that is growing faster than the development of constraining international norms, increasing the prospects for miscalculations and escalation.[2] This book explains how constraining norms for emerging-technology weapons, including cyber weapons, will develop. It does so by expanding general norm evolution theory to better apply to emerging-technology weapons based on case studies on norms developed for chemical, biological, and nuclear weapons and strategic bombing. As a validation of norm evolution theory, elements from nearly every general hypothesis theory were present in one or more of the three case studies, and many were in all three. However, general norm evolution theory is an imprecise tool for predicting norm evolution for emerging-technology weapons and was largely developed for other issues. As such, a tailored version of norm evolution theory was developed, which offers the primary hypothesis that direct or indirect alignment of national self-interest with a constraining norm leads to norm emergence, and the extent to which it is aligned with key or powerful states' perception of self-interest will determine how rapidly and effectively the norm emerges. This new norm evolution theory for emerging-technology weapons suggests certain secondary variables and hypotheses regarding actors, actor motives, and important mechanisms and factors that influence norm emergence and growth during each stage of the norm life cycle. Applying this analytic framework to a plausible future cyber warfare scenario provides insight into how constraining cyber norms will develop in the coming years. There are currently multiple candidate norms for cyber warfare that are beginning to emerge, such as a total prohibition on cyber warfare, a prohibition on first use of cyber weapons, and the application of the existing LOAC to cyber warfare. These candidate norms are being developed through state practice and deliberate norm cultivation efforts through an increasing number of organizational platforms, which norm evolution theory indicates is a sign of progress. However, when the analytic framework for norm evolution for emerging-technology weapons is applied, it appears that constraining norms for cyber warfare will have trouble emerging and may not ever reach a norm cascade. This is due to a variety of factors, primarily the apparent lack of a convergence of powerful self-interested actors perceiving the norms as useful due to differing perspectives on the impact of cyber capabilities because of limited demonstrations and the diffusion of cyber capabilities. Should the constraining norms emerge successfully, the secondary variables in this new framework suggest that the odds that these norms will reach a tipping point are better. However, internalization is less likely. Potential contingency scenarios, such as the occurrence of a major cyber attack, could alter the prospects of cyber warfare norm evolution.

In light of this conclusion, I offer several recommendations for U.S. policymakers to consider as they grapple with the threat of cyber warfare. These recommendations are based on a realist perspective that views the international system as fundamentally anarchic, albeit with nonmaterial normative factors playing a role, as well as the belief that states consistently pursue their own interest, defined as power. Based on this perspective, an operating assumption is that U.S. primacy is desirable and that U.S. policy should pursue the goal of preserving its primacy by achieving the best net outcome in cyberspace.

1. Pursue Only Limited Consensus on Cyber Norms and Do Not Overly Constrain Development of Offensive Cyber Capabilities

Norms for the use of force in cyberspace are very much in flux. Some in the United States have advocated for the aggressive pursuit of various types of norms, leading up to ambitious ideas regarding "cyber arms control." Adm. Bill Owens, former vice chairman of the Joint Chiefs of Staff, has pushed for an agreement on "no first use of cyber-attack," and Richard Clarke, former national coordinator for security, infrastructure protection, and counter-terrorism for the United States and special advisor to the president on cybersecurity, has argued for a "Cyber War Limitation Treaty."[3] However, this book has shown that the prospects are not good for the successful emergence and development of such constraining norms.

Pursuing comprehensive constraining norms as a norm leader necessarily requires some constraints on U.S. cyber weapons because developing and demonstrating these capabilities while also attempting to establish norms would be perceived as hypocritical and counterproductive. This tension was highlighted by the purported U.S. decision to launch Olympic Games and the Stuxnet attack. Referring to the attack, Jason Healey, director of the Atlantic Council's Cyber Statecraft Initiative, said, "If we were in such danger with SCADA, should we have thrown the first SCADA punch?," and former CIA and NSA director Michael V. Hayden said, "In a time of peace, someone just used a cyber weapon to destroy another nation's critical infrastructure . . . ouch."[4]

Clearly, U.S. use of cyber warfare capabilities and "militarization" of cyberspace has negative effects on norm emergence given the reciprocal nature of fragile nascent norms, as evidenced by Iran's retaliatory attacks on Saudi Aramco.[5] However, in light of the tremendous capabilities offered by cyber weapons and the long odds for the successful evolution of constraining norms, the U.S. should not dramatically constrain further development of its offensive cyber capabilities and should instead pursue only limited, modest norms. The aggressive pursuit of constraining norms may not be worth the investment and trade-offs involved: the United States should keep its options open and prepare for major cyber war, which is only marginally constrained by norms. This is not to say that the United States should aggressively pursue and prepare for unconstrained cyber warfare, only that it should not forgo offensive cyber capabilities in the hope that constraining norms will mitigate the cyber threat posed by other nations. In the future, circumstances may change, and this strategic decision could be revisited. The trend of U.S. investment in cyber weapons indicates that this recommended path is likely already being pursued. The 2013 National Defense Authorization Act (NDAA) stressed that "the advantage is on the offense" regarding cyber warfare and authorized the development of offensive cyber capabilities.[6] The 2014 NDAA increases the funding authorized for these capabilities in addition to establishing some requirements to attempt to limit the international proliferation of cyber weapons.[7]

2. Judiciously Employ Offensive Cyber Capabilities, Perhaps Offering a Cyber Extended Deterrent to Allies to Manage Proliferation

While the United States should pursue only limited consensus on constraining cyber norms and not constrain its development of offensive cyber capabilities, there is still some hope for cyber norms playing a stabilizing and constraining role in international security. Therefore, norms should not be actively undermined by unnecessarily promiscuous use of cyber capabilities. Instead, these capabilities should be employed judiciously. Ralph Langner, in his detailed analysis of Stuxnet, argues that another purpose of the advanced counterproliferation cyber weapon was to demonstrate that cyber weapons work—the equivalent of a *Sputnik* moment for cyber warfare.[8] This has certainly been achieved, so future cyber attacks for demonstration purposes are unnecessary. Cyber capabilities should be rare and limited to certain targets, such as military targets (which is consistent with recent U.S. cyber attacks). International relations scholar John Arquilla outlined some potential applications of limited yet effective use of cyber warfare in his controversial 2009 comments, arguing that the United States should go on a limited "cyber offensive."[9] He suggested the selective use of cyber warfare as a "nonlethal way to deter lethal conflict" and outlined scenarios such as a looming Pakistani-Indian or Russian-Georgian conflict where such capabilities could be used to potentially avert a kinetic conflict.

This limited use would not require a deviation from current practices but would prevent further escalation and unnecessarily undermine norm evolution. Healey has pointed out that large, powerful nations generally have not carried out large-scale cyber attacks outside of existing conflicts, even against small nations.[10] For example, Stuxnet was employed only in the context of the existing conflict between the United States/Israel and Iran, and Estonia was attacked in the context of an existing conflict with Russia. Robust U.S. development of offensive attack capabilities could also end up providing space for constraining norms to grow by offering "cyber extended deterrence" to allies, thus limiting the proliferation of sophisticated capabilities. This extended deterrence could be similar to extended nuclear deterrence offered to U.S. allies who lack nuclear weapons. Cyber weapons may not be suitable for achieving direct deterrence due to their unpredictability and unreliable effects, but generally offering cyber capabilities to allies or committing to conducting operations on their behalf under certain circumstances could help them determine to limit their own cyber warfare investments. This dynamic is already at play. As Japan looks to establish a "cyber defense task force" and develop its own offensive cyber capabilities, it recognizes it is reliant on U.S. cooperation and support.[11] Likewise, Australia is anticipating a benefit from the United Kingdom's investment in offensive cyber capabilities.[12]

3. Seek Limited Norm Cultivation

The United States should actively pursue more limited constraining norms for the use of cyber weapons rather than norms constraining their proliferation and development. Of the current candidate cyber norms identified in chapter 5, the most likely to succeed are those that are more limited, such as those focused on the application of the existing LOAC to cyber warfare.[13] Perhaps the most viable limited constraining norm for the near and mid-term would be the application of this noncombatant immunity norm specifically to IT systems associated with the financial industry due to the global economic interdependence and the potential for economic chaos should trust in these systems be shaken. Such a norm is likely to most readily generate a convergence of self-interested support from key states as well as congruent private sector support—factors norm evolution theory indicates is important as applied to emerging-technology weapons. The United States could pursue this limited constraining norm by leveraging its unique and powerful role as the location of most globally desired Internet services. Tools such as isolation of violating countries through Internet service provider blockades could compel state support for the norm and force states to exercise responsibility for nonstate violations originating from their territory.

4. Manage Potential Miscalculation with Increased Transparency and Confidence-Building Measures

The United States should attempt to address the shroud of government secrecy in cyberspace through increased transparency and use of CBMs. Given the challenge of attribution, intense secrecy, and the ambiguous relationship between cyber war and cyber espionage, the prospect for miscalculation and inadvertent escalation is heightened.[14] To help address these issues and minimize this risk, the United States should embrace the voluntary CBMs developed by the UN GGE and ICT4Peace identified in chapter 5. These include measures such as exchanging views and information on national strategies, policies, and organizations as well as the creation of bilateral and multilateral consultative frameworks such as seminars and workshops to discuss and foster a better understanding of each nation's approach to cyber warfare.[15] The United States should also engage in bilateral dialogue with other key cyber actors to achieve a consensus on key terms, communicate red lines, and develop crisis communication mechanisms. This could be similar to the consultations the EastWest Institute has organized to discuss cyber diplomacy, resulting in a report outlining how Russia and the United States can work together to find common ground on norms for cyber warfare (among other things).[16] The United States should also seek to establish additional crisis communication hotlines, similar to the one it recently established with Russia.[17] These efforts will help to prevent miscalculation and could help create more trust and consensus in order to bring about the development of comprehensive constraining norms in the years ahead.

Many nations are incorporating cyber warfare capabilities into their national security organizations. Some have likened this ongoing cyber revolution—perhaps with too much hyperbole—as the biggest military revolution since tanks replaced cavalry nearly a hundred years ago.[18] Along these lines, Iran's supreme leader Ayatollah Ali Khamenei, perhaps reflecting the lessons his country learned on the receiving end of Stuxnet, said in early 2014, "Cyber, in my modest opinion, will soon be revealed to be the biggest revolution in warfare, more than gunpowder and the utilization of air power in the last century."[19] While the norm evolution theory for emerging-technology weapons developed in this book indicates that there is limited hope of rapidly developing constraining international norms to contain this threat, U.S. national security may eventually be enhanced by the emergence of some kind of regulative norm for cyber warfare, similar to those that developed in the past for other emerging-technology weapons. In the meantime, pursuing the four recommendations above will best serve U.S. national security interests.

Additional research is warranted to further develop norm evolution theory for emerging-technology weapons, as such a theory has broad implications for norm evolution across a range of new weapon types as we head deeper into the twenty-first century. Autonomous weapon systems, high-energy lasers, microwave-based active denial systems, electromagnetic rail guns, and quantum stealth technology are all emerging-technology weapons for which norms are in the earliest stages of development.[20] Further validating and enhancing norm evolution theory for this category of weapons will provide greater insight into how constraining or permissive norms may evolve and what, if anything, policymakers can do to shape their development so as to protect their security interests. Accordingly, in addition to the four recommendations for U.S. policymakers listed above, researchers should seek to further validate and refine norm evolution theory for emerging-technology weapons. To do so, this book recommends an agenda of additional historic case studies of emerging-technology weapons in order to further develop the framework. These could include the evolution of norms for (1) the use of gunpowder weapons in the fifteenth through seventeenth centuries, (2) submarine and anti-submarine warfare in the early and mid-twentieth century, (3) the use of remotely piloted aircraft (also known as unmanned aerial vehicles or drones) in the late twentieth and early twenty-first centuries, and (4) efforts to achieve the bioenhancement of soldiers today and in the recent past. In addition to developing these historic emerging-technology weapon case studies to continue to better understand how norms for emerging-technology weapons evolve, researchers should also seek better mechanisms to measure the empirical impact of norms for war. Current norm evolution theory lacks a well-developed means to determine the relative impact of international norms on state behavior.[21] Empirically measuring the impact of norms is possible only when the behavior observed is compatible with the norm and simultaneously incompatible with the clear rational state interest.

If the two are aligned, it is extremely difficult to measure the extent to which norms played a role. Even when norms and state interest are in conflict and the path of the norm is chosen, other variables may be at play. To fully measure and attribute the behavior to the influence of the norm, evidence needs to exist citing the existence of the norm as a part of the actor's decision-making calculus. All this is to say that measuring the impact of norms rather than how they develop and evolve, which is the focus of this book, is particularly difficult, and further research should seek to better address this issue. Such an effort could be pursued by historic case studies of state decision making involving weapons and warfare in specific cases where norms and state interest were aligned and in conflict in order to develop an approach to evaluate their relative impact.

APPENDIX

Background on Cyber Warfare

The cyberspace domain is defined in numerous ways and has only recently emerged as a strategic security concern. Understanding this domain is essential before one can consider how norms for cyber warfare will develop. In the United States the domain was originally defined by DOD in 2000 as the "notional environment in which digitized information is communicated over computer networks."[1] This computer-centric definition was significantly modified in 2006 when the U.S. Air Force offered a broader definition that was subsequently adopted by the Joint Chiefs of Staff in late 2006 and ultimately codified for all of DOD.[2] The new military definition of cyberspace, which applies to the military and nonmilitary sectors, is "a global domain within the information environment consisting of the interdependent network of information technology infrastructures, including the Internet, telecommunications networks, computer systems, and embedded processes and controllers."[3]

This definition encompasses the Internet, the World Wide Web, smartphones, computer servers, iPads, and other common resources. The U.S. government's 2003 "National Strategy to Secure Cyberspace" specified a list of sectors particularly reliant on cyberspace: agriculture, food, water, public health, emergency services, government, defense industrial base, information and telecommunications, energy, transportation, banking and finance, chemicals and hazardous materials, and postal and shipping.[4] Given the breadth of functions of daily life reflected in this list, cyberspace is unmistakably central to the U.S. and global economy. The United States alone has over 239 million regular Internet users (a 77.3 percent penetration rate).[5] Cyberspace is also a key supporting element of U.S. military power. The DOD relies heavily on IT networks for command, control, communications, computer, intelligence, surveillance, and reconnaissance and the planning and execution of day-to-day military operations. This reliance on cyberspace also applies to the rest of the international community. As the Obama administration's "International Strategy for Cyberspace" points out, "The last two decades have seen the swift and unprecedented growth of the Internet as a social medium; the growing reliance of societies on networked information systems to control critical infrastructures and communications systems essential to modern life; and increasing evidence that governments are seeking to exercise traditional national power through cyberspace."[6]

The ITU, the UN agency for information and communication technologies, reported that over one third of the world's seven billion people were online at the end of 2011, a 17 percent increase since 2006.[7] Multilateral security organizations such as NATO are still grappling with how to approach cyber threats and develop consensus on regulative norms and approaches for collective defense.[8] The cyberspace domain is largely owned and controlled by private industry, and thus many actions in cyberspace require a public-private partnership.[9] This raises a host of ethical and legal questions associated with conducting warfare through a domain largely privately owned and controlled. What are the responsibilities of Internet service providers (ISPs) to detect, report, and block malicious traffic intended to harm their host nations? This legal question and many others arising from this rather unique aspect of the domain have yet to be resolved.

While it can be challenging to reach agreement on what constitutes cyberspace as a domain, hostile action in cyberspace is even more difficult to define, yet it is crucial to understand the dynamics of cyber warfare before examining the prospect of norm evolution for emerging-technology cyber weapons. Cyberspace operations are the employment of cyber capabilities where the primary purpose is to achieve objectives in or through cyberspace and cyber warfare and generally understood to be CNE- and CNA-style attacks. Often CNE and CNA go hand in hand as CNE is conducted to collect information and conduct reconnaissance prior to a CNA. The line between these two major categories of hostile action in cyberspace is often blurred. In a very real sense, using unauthorized cyber access to steal information allows the option to destroy information and progress into a cyber attack. Journalist Tom Gjelten described this phenomenon: "The difference between cyber crime, cyber-espionage and cyber war is a couple of keystrokes. The same technique that gets you in to steal money, patented blueprint information, or chemical formulas is the same technique that a nation-state would use to get in and destroy things."[10] As a result many today refer to cyber espionage as "cyber warfare" or "cyber attacks" when in actuality no damage (other than secondary damage caused by the relative advantage the stolen information provides) occurs. The cyber theorist John Arquilla points out that international law defines an attack as "violence against the adversary" and that such a term does not necessarily apply to all cyber operations (namely, CNE).[11] As an example of this blurred line, in 2014 a cyber attack occurred during the political crisis in Ukraine, involving a weapon known as Snake, or Ouroboros. Snake is of suspected Russian origin, although positive attribution has not been achieved.[12] It is a CNE, or possibly CNA, tool kit that in 2010 began infecting Ukrainian computer systems.[13] Since 2010 researchers have identified fifty-six incidents of Snake, thirty-two of them in Ukraine, and believe it was used not only for CNE but also to conduct highly sophisticated CNA-style attacks.[14] This imprecise lexicon of *cyber warfare* and *cyber attack* complicate the social environment in which norms for actual cyber warfare must emerge and develop. Security and Defence Agenda, in collaboration with the computer security company McAfee, published a report in February 2012 identifying the lack of agreement over key terms such as *cyber war* and *cyber attack* as a major impediment to norms and regulating cyber conflict.[15]

Cyber warfare represents a major RMA. Some have gone so far as to predict that it will "soon be revealed to be the biggest revolution in warfare, more than gunpowder and the utilization of air power in the last century."[16] In all likelihood, the threat of emerging-technology CNA-style cyber weapons will only increase. CSIS has identified more than thirty countries that are taking steps to incorporate cyber warfare capabilities into their military planning and organizations, and Adam Liff has argued that the use of cyber warfare as a "brute force" weapon is likely to increase in frequency.[17] China uses the term *informationized wars*, which are "heavily reliant on computers and information systems and focus on attacking such systems possessed by their adversaries."[18] Increased international interest in cyber warfare is also based on the recognition that information networks in cyberspace are becoming operational centers of gravity in armed conflict.[19] This was reflected in DOD's 2014 Quadrennial Defense Review (as it was in previous reviews):

> The United States has come to depend on cyberspace to communicate in new ways, to make and store wealth, to deliver essential

> services, and to perform national security functions. The importance of cyberspace to the American way of life—and to the Nation's security—makes cyberspace an attractive target for those seeking to challenge our security and economic order. Cyberspace will continue to feature increasing opportunities but also constant conflict and competition—with vulnerabilities continually being created with changes in hardware, software, network configurations, and patterns of human use.[20]

Cyber warfare plays a role at the tactical, operational, and strategic levels of war: impacting engagement systems at the tactical level, the adversary's ability to mass and synchronize forces at the operational level, and the ability of senior leadership to maintain clear situational awareness of the national security environment at the strategic level.[21] Scholar Michael Horowitz's theory on the diffusion of new military capabilities, known as adoption capacity theory, argues that cyber weapons are likely to spread quickly and that the diffusion of military innovations depends on two intervening variables: the financial intensity involved in adopting the capability and the internal organizational capacity to accommodate any necessary changes in recruiting, training, or operations to adopt the capability.[22] The low financial and organizational barriers to developing cyber warfare capabilities suggest that the adoption of cyber warfare will likely be widespread.

Cyber warfare has many characteristics that often do not apply to other forms of conflict, especially conventional military conflict. These include the challenges of actor attribution, the multiuse nature of the associated technologies, target and weapon unpredictability, the potential for major collateral damage or unintended consequences due to cyberspace's "borderless" domain, questionable deterrence value, the use of covert programs for development, attractiveness to weaker powers and nonstate actors as an asymmetric weapon, and use as a force multiplier for conventional military operations.[23] It is necessary to consider these factors when evaluating the prospects of norm emergence and development for cyber warfare as they may make fostering certain types of norms more or less difficult.

Challenge of Attribution

The first major characteristic of most cyber weapons is the challenge of attribution following their use. This is a result of the tremendous difficulty in conclusively determining the origin, identity, and intent of an actor operating in this domain if the actor wishes to remain anonymous, and defenders generally lack the tools needed to trace an attack back to the actual attacker. Cyber scholar Thomas Rid argues that all cyber attacks to date have been sophisticated forms of sabotage, espionage, or subversion and are reliant on this attribution difficulty.[24] Cyberspace is truly global; nearly all action passes through networks and ISPs in multiple countries. Additionally, the hardware used to conduct cyber warfare can be owned by innocent noncombatants, illicitly harnessed for malicious use through viruses (as was the case in the Estonian and Georgian attacks). Some computer experts estimate that between 10 and 25 percent of computers connected to the Internet (approximately 100 million to 150 million devices) are compromised and used illicitly as part of various networks of compromised computers—known as "botnets"—utilized to conduct attacks.[25] The use of these proxies provides plausible deniability to state-sponsored activity. The Conficker worm, first detected in November 2008, is another example of the challenge of attribution in cyberspace. It is suspected to be of Ukrainian origin, largely because it did not target Ukrainian Internet protocol (IP) addresses or computers using Ukrainian-configured keyboards; however a savvy adversary could have deliberately programmed it that way as part of a deception strategy.[26] Another attack, this one on DOD computer systems and known as Solar Sunrise, was initially traced back to Israel and the United Arab Emirates. U.S. officials suspected that the attack was orchestrated by operatives in Iraq; later investigations determined it was conducted by two teenagers in California.[27] Yet another cyber attack, Night Dragon, targeted five multinational oil companies and stole gigabytes of highly sensitive commercial information on Western energy development activities. Investigators traced the attack to IP addresses in China and confirmed that the tools used in the attack were largely of Chinese origin and that the attacks were conducted between 9:00 a.m. and 5:00 p.m. Beijing time, indicating the likelihood that government or government-affiliated personnel conducted the attack. In spite of this significant evidence indicating Chinese involvement, probably even official Chinese government involvement, it was not possible to conclusively attribute the attack, and Chinese officials claimed innocence.[28] Attribution challenges make cyber warfare particularly appealing to an adversary seeking to achieve certain effects anonymously or at least with reasonable deniability.

Multiuse Nature of Technology

Another characteristic of cyber weapons is that their underlying technologies are multiuse. This means that cyber IT systems can have defensive and civilian applications and purposes in addition to any offensive cyber warfare application. In fact many IT and hardware and software components usable for cyber warfare are ubiquitous commercial off-the-shelf technologies with many peaceful applications. According to the National Research Council (NRC), advances in IT are driven primarily by commercial needs and are widely available across the globe to nearly all groups and individuals.[29] Forrester Research projects that the number of computers worldwide, and therefore the number of individuals with access to these tools, will grow from one billion in 2008 to two billion by 2015.[30] The military and intelligence community IT required to conduct cyber warfare is drawn from these globally distributed and commercially developed resources. In some cyber operations, such as those utilizing distributed denial of service (DDOS) attacks, private and commercial computers themselves may deliberately and surreptitiously be utilized. A DDOS attack uses multiple compromised systems (collectively known as botnets), usually infected with a Trojan virus that can be developed by simple criminals or state actors, to target a single system. Victims of a DDOS attack are the end-targeted system and all systems maliciously used and controlled by the attacker (also known as a "botherder" or "botmaster") in the distributed attack. The most common form of DDOS attack is simply to send more traffic to a network address than it is equipped to handle. This multiuse nature of cyber warfare technology has obvious implications for the ability to address cyber threats by restricting access to the hardware or software involved; doing so would likely not be particularly effective or practical.

Unpredictability and Potential for Collateral Damage

Another characteristic of cyber weapons is the unpredictability and potential for collateral damage associated with their use. Due to the ever-changing innovations in enterprise architecture and network operations, as well as any IT interdependencies, predicting the precise effects of an attack are very difficult. As in other war-fighting domains, an actor may have conducted intelligence, surveillance, and reconnaissance (ISR) operations and mapped out vulnerabilities in an adversary's cyber network, as would be done to plan for a conventional ground attack with tanks and troops. However, unlike in the conventional realm, the targeted actor is capable of flipping a switch and instantly changing the network (that is, the target set) or even unplugging it altogether. This factor is a destabilizing force as it rewards immediate hostile action to prevent network modification if cyber ISR intrusions are later detected. It is in effect the opposite of deterrence, incentivizing early offensive strikes when an advantage is present. Defenders may also have unknown automated countermeasures that negate the desired effects of cyber attacks (such as instantaneous network reconfiguration or firewalls). For example, the Stuxnet attack is likely no longer able to continue to attack Iranian nuclear facilities as the zero-day exploits it utilized have been plugged by Iranian officials. In addition to network and target evolution, cyber weapons themselves can also be unpredictable since many can evolve. A cyber weapon can adapt, as has been seen with the Conficker virus. Conficker includes a mechanism that utilizes a randomizing function to generate a new list of 250 domain names (used as command and control rendezvous points) on a daily basis, remaining adaptable and staying one step ahead of those seeking to shut down or hijack the illicit Conficker-enabled network.[31]

Network interdependencies also contribute to the potential for collateral damage that is characteristic of cyber weapons. The Internet is made up of hundreds of millions of computers connected through an elaborate and organic interwoven network and is the backbone of much of the global economy; any major attack could pose significant unintended and collateral impacts if it spurred a ripple effect through the network. For example, if an attack on a particular Internet node resulted in the blackout of an entire regional ISP, not only would the intended target be affected but also all other users of the ISP and other individuals who relied on the services of those users directly impacted. The second- and third-degree impacts of some forms of cyber warfare are nearly impossible to predict. These effects are not limited to the theoretical: cyber attacks have already led to real-world collateral damage. Israel's suspected cyber attack on Syrian air defense radar in advance of their 2007 attack on a Syrian nuclear reactor under construction may have also inadvertently caused damage to Israel's own cyber networks.[32]

Questionable Deterrent Value

Many cyber weapons have questionable value in achieving deterrent effects. The uncertain effects of cyber weapons coupled with the availability of defenses and the need for secrecy and surprise reduce their ability to serve as a strategic deterrent. Available defenses and the potential for network evolution to mitigate the effects of an attack given early warning require cyber attackers to rely on surprise for much of their effectiveness. To achieve surprise, secrecy is required, reducing the ability of a state to make credible threats without compromising their cyber warfare capabilities. Credible threats regarding specific means of attack or targets invite the threatened state to take protective actions, which could blunt the deterrent value of a threat. While cyber weapons can cause significant, perhaps unacceptable damage, they have difficulty doing so in an "assured" or reliable manner, which consequently undermines their utility for deterring adversaries. Additionally, because of the attribution challenge, there is limited public discussion regarding cyber warfare capabilities and intent. The secrecy surrounding cyber programs, noted in a 2011 CSIS report, indicates that at this time states likely do not plan on using cyber capabilities to achieve deterrence effects.[33]

Importance of Secrecy and Surprise

Due to the sensitivities of cyber weapons and the uncertain international response, their development is rarely publicly acknowledged or demonstrated, and because of the multiuse nature of IT, the development of offensive cyber capabilities is similar in many ways to the development of defensive capabilities or even civilian and commercial activities. Thus it can be very difficult to gauge the intentions of an adversary based solely on public indicators. Cyber warfare does not require large facilities with distinctive signatures and easily detected emissions, as would a nuclear weapons program. This makes national technical means such as intelligence collection satellites fairly ineffective for understanding adversary cyber activities. Intelligence on foreign capabilities and intentions in both areas is likely to be poor barring well-placed human sources, who pose challenges of their own.[34] The utility of cyber CNE-type espionage activity incentivizes keeping efforts to develop such cyber tools, and countermeasures against them, secret. There is very limited public discussion regarding national cyber capabilities and planning. CSIS reported that many states keep information about their cyber warfare programs and capabilities secret.[35]

Secrecy and surprise in cyberspace may also result in the emergence of revolutionary technology such as a quantum computer, which would render all forms of encryption obsolete. Any category of weapon is always subject to an advance in technology that gives someone an edge, but with cyber the risk of a breakout is much more pronounced. A quantum computer would utilize the principles of quantum mechanical phenomena to process data at spectacular speeds. This technology is no longer just theoretical; in 2007 D-Wave Systems, a Canadian corporation, produced an extremely basic 16-qubit quantum computer, and recently Lockheed Martin began operating a 512-qubit machine.[36] The first nation to develop and field a full-blown quantum computer will be able to utterly dominate cyberspace for a period of time. Looking beyond the acute example of quantum computing, smaller technological advances could also have a dramatic effect on the balance of power in cyberspace. The life cycle of advanced computer technology is much more accelerated than other weapon systems. Moore's Law, developed by Intel's cofounder Gordon Moore in the 1960s, rather accurately predicted that computer technology would advance dramatically: "The number of transistors which can be manufactured on a single die will double every 18 months."[37] Moore's Law continues to apply today. If a nation fails to keep up with these advances, its ability to defend against or wage cyber warfare will be dramatically reduced.

Asymmetric Warfare

Cyber weapons are attractive to weaker powers and nonstate actors as asymmetric weapons. This attractiveness is based on the potential for anonymity and associated plausible deniability as well as the relatively low cost of development and the global power projection cyber weapons can provide. The most successful known cyber weapon, Stuxnet, likely cost in the low double-digit millions of dollars to produce.[38] Alternative weapons for achieving similar effects against the Iranian nuclear program, Stuxnet's target, would have necessitated weapons costing billions of dollars (for example, producing a single B2 bomber costs over $2 billion).[39] Cyber expert Adam Liff has contested the financial ease of acquiring potent cyber weapons and argued that obtaining advanced cyber weapons such as Stuxnet would in most cases exceed the reach of weaker states.[40] However, Liff does not take into account the ease with which computer code, once developed, can be replicated and modified. Eugene Kaspersky, the founder of an antivirus software company that bears his name, has said that given that Stuxnet's code is now publicly available, it would be "quite easy to disassemble the code to discover how it works, to extract the components and to redesign the same idea in a different way."[41] As a result the cost of cyber weapons will likely decrease as they and their associated code proliferate and are increasingly deployed. Information security researcher Dorothy Denning described this appeal to weaker actors when she highlighted that the cost of launching cyber warfare operations could be "negligible," while the cost to the attack's victims could be "immeasurable."[42]

In addition to relative low cost, cyber weapons also provide global power projection capability to almost any adversary due to the global nature of cyberspace. This characteristic is particularly appealing to states with very limited expeditionary capabilities but with global aspirations, such as China. "Thanks to computers," one Chinese strategist writes, "long-distance surveillance and accurate, powerful and long-distance attacks are now available to our military."[43] Operations in cyberspace, unlike those in other domains (with the possible exception of space), immediately give a state global power projection capability. The NRC has highlighted this prospect of "remote-access" attacks, where computers are attacked through the Internet or connection nodes present in wireless networks or dial-up modems.[44] By tapping into global ISPs and other IT-based networks, attackers are able to effectively conduct expeditionary warfare in an area distant from their own territory. Prior to the advent of cyber warfare, very few nations had the resources to develop the sizable and robust military assets required to overcome global logistical challenges and project power far outside of their neighborhood. Through preexisting global computer networks, a cyber attack with global reach can be conducted as rapidly as electrons can traverse the electromagnetic spectrum. While IT networks and advanced technologies have enhanced the command and control required for traditional power projection, cyber attacks can now be conducted from completely within cyberspace itself. This effectively removes the high entry costs required in conventional warfare to develop aircraft carrier battle groups, strategic bombers, intercontinental ballistic missiles, and so forth, associated with power projection. Denning writes that cyber warfare operations "can take place in an instant and come from anywhere in the world. They can be orchestrated and conducted from the comfort of a home or office, without the risks of spies and undercover operations, physical break-ins, and the handling of explosives. The number of targets that potentially could be reached is staggering."[45]

Cyber warfare has clear limitations compared to traditional expeditionary capabilities, but it is understandably attractive to less developed or advanced states, such as China and other rising peer competitors to the United States who are seeking to exert global influence. The 2007 DDOS attacks against Estonia are a good example of this power projection capability. During a two-week period attackers were able to successfully disrupt the Estonian government, media outlets, banking, ISPs, and telecommunications websites by launching attacks from approximately 100 million computers distributed in more than fifty nations.[46] Due to the asymmetric nature of cyber warfare, it is likely to be a favored form of warfare by adversaries unwilling to directly challenge conventional military capabilities with similar conventional capabilities (particularly China, which has demonstrated a heightened interest in cyber warfare).[47]

Force Multiplier

The ability to use cyber weapons as a force multiplier for conventional military operations is another significant characteristic of cyber warfare. Cyber weapons are well suited for attacks on logistical networks, reinforcements, and command-and-control facilities "to induce operational paralysis, which reduces the enemy's ability to move and coordinate forces in the theater."[48] While cyber weapons may not have a direct kinetic effect on an adversary's tanks and aircraft, it is still possible for cyber attacks to render these weapons useless. Because cyber weapons can achieve such effects without kinetic destruction, they can be employed in ways similar to those intended for the infamous neutron bomb, which killed troops with a blast of lethal neutron radiation but did not damage buildings or physical infrastructure. Thus cyber weapons can provide an attacker with the capability of seizing valuable natural resources or industrial facilities without risking their destruction.

Similarly cyber weapons, particularly those allegedly being developed by China to exploit the U.S. military's logistics IT network, would complement conventional military operations. A 2007 RAND Corporation report on Chinese anti-access strategies explained that Chinese military strategists believe cyber attacks are likely to be effective in disrupting U.S. military operations because military IT systems are connected to commercial networks. One Chinese official pointed out that in the United States, "95 percent of military networks pass through civilian lines and that 150,000 military computers pass through normal computer networks. This characteristic of computer networks makes it easy to conduct a virus attack."[49]

Despite their general lack of transparency on defense issues, Chinese strategists have had a handful of open discussions about how they would exploit this weakness as a force multiplier for a conventional conflict. A Chinese report in 2000 stated that the goal of Chinese cyber warfare was to "cut off the enemy's ability to obtain, control, and use information, to influence, reduce, and even destroy the enemy's capabilities of observing, decision-making, and commanding and controlling troops, while we maintain our own ability to command and control."[50]

State Practice of Cyber Warfare Today

A major mechanism that facilities the emergence of international norms and ultimately international law as codified in treaties and conventions is the concept of customary practice leading to customary international law. Customary international law "develops from the general and consistent practice of states if the practice is followed out of a sense of legal obligation."[51] In this early stage of the cyber era, norms for cyber warfare emerge in part based on this "general and consistent" practice of states just as much as it arises from deliberate efforts and diplomatic dialogue. There is no precise formula regarding how long a practice must occur before it becomes an international norm or customary international law, but the practice of major powers obviously has more significance than less prominent or powerful states.

Indicators of Early Norms

While most hostile cyber operations to date can be properly classified as CNE, some attacks provide insight into the emerging customary practice of states and related emerging norms in regard to this most serious type of hostile cyber operation. Consciously or unconsciously, early cyber actors are acting as the early norm leaders identified by norm evolution theory as they help establish customary practice for hostile operations in cyberspace. Many small CNA-style operations involve DDOS attacks to degrade access to websites, such as the Code Red attack in 2001, which involved malware that launched a DDOS attack against White House computers.[52] It is believed that approximately 100 million to 150 million botnets are utilized to conduct these frequent DDOS attacks.[53] However, there are few examples of major cyber warfare attacks. Seven will be examined for the purposes of evaluating the norms and contemporary state practice of cyber warfare: the purported attacks on a Siberian gas pipeline in 1982, the DDOS attacks on Estonia in 2007, the Israeli Operation Orchard attacks on Syria in 2007, the attacks on Georgia in 2008, the notorious Stuxnet attack on Iran disclosed in 2010, the Shamoon virus attack on Saudi Aramco in 2012, and Izz ad-Din al-Qassam's Operation Ababil attack against financial institutions in 2012. This discussion focuses on the nature and target of the attack in order to better understand the current state of norms for cyber warfare. It does not provide a comprehensive or detailed technical examination of each attack.

The Trans-Siberia Pipeline Attack

While cyberspace as we know it today has existed for only two decades and most sophisticated cyber attacks have occurred only in the past decade, the first purported CNA-style cyber operation dates back to 1982. This attack is largely still shrouded in uncertainty. In 1982 a portion of the Trans-Siberia pipeline within the Soviet Union exploded, allegedly as a result of computer malware implanted in the pirated Canadian software by the CIA, which caused the SCADA system that ran the pipeline to malfunction.[54] The main source of information on this cyber attack is the "Farewell Dossier," which states that "contrived computer chips (would make) their way into Soviet military equipment, flawed turbines were installed on a gas pipeline, and defective plans disrupted the output of chemical plants and a tractor factory."[55] While the accuracy of this attack is disputed to this day, it allegedly resulted in the "most monumental nonnuclear explosion and fire ever seen from space"; the embarrassed Soviets never accused the United States of the attack.[56] For the purposes of understanding emerging norms, this event is significant because it involved an attack on critical infrastructure that was not explicitly military in nature. The Trans-Siberian pipeline was responsible for transporting natural gas to western Ukraine and ultimately to the broader energy market and generated revenue of about $8 billion a year.[57] The attack demonstrates that constraining norms limiting cyber warfare's use against nonmilitary targets did not exist in 1982. More recent examples indicate that perhaps these constraining norms still do not exist.

The Estonia Attack

The cyber attacks against Estonia are a more recent example of states' emerging contemporary practice of cyber warfare. In late April 2007 the Estonian government's efforts to relocate a Soviet-era statue in their capital city of Tallinn led to significant disruptions on their Internet and web-based services that lasted for several weeks and consisted of 128 unique DDOS cyber warfare attacks. At its peak, traffic originating from outside Estonia was four hundred times higher than its normal rate and involved approximately 100 million computers in more than fifty countries—highlighting some of the issues associated with the idea of discrimination and noncombatant immunity. The attackers executed the attacks using a series of botnets, and investigators determined that the attacks were carefully coordinated in advance due to the fact that the attack did not propagate and did not appear to be centrally controlled through an identifiable command-and-control center.[58] To alleviate the attacks, Estonian telecommunications companies and ISPs worked quickly to expand network capacity and move government sites to alternate servers. The cyber warfare waged against Estonia was the first time a sophisticated attack focused on disruption and denial of services was conducted against a nation-state. Many sources believe the Russian government was involved due to the large number of IP addresses originating in Russia, as well as the obvious motive for their engagement. Because of the cyber attribution challenge, no "smoking gun" evidence has been made public to support that notion and the Russian Federation has denied any involvement. Estonian officials have been unable to identify and apprehend the perpetrators. Following the attack, NATO, of which Estonia is a member, established the NATO CCD COE on May 14, 2008.[59] This center, located in Tallinn, seeks to enhance NATO's ability to respond to cyber attacks and as of late has been acting as an organizational platform for norm entrepreneurs. The Estonia cyber attacks were aimed directly at disrupting and degrading civilian services and thus demonstrated the lack of a constraining cyber norm for noncombatant immunity or discrimination. However, the attacks did not result in permanent damage and did not destroy any critical infrastructure—although this was likely due to the limits of the DDOS mechanism available and not to any normative constraint.

Operation Orchard

One of the first examples of cyber warfare that was designed to support—though not directly cause—real-world physical damage was Israel's suspected cyber attacks on Syrian air defense radar in advance of their 2007 attack on a Syrian nuclear reactor under construction. Known as Operation Orchard, the attack is believed to have caused meaningful degradation of Syria's air defenses and thus helped enable the Israeli aircraft to cause the physical destruction of the Syrian nuclear site. This attack targeted a military target in support of an attack on another military objective. Syria did not protest the cyber attack, as doing so would have required acknowledging its illicit nuclear program. Of further interest to constraining norms for cyber warfare, it is believed that the Israeli offensive cyber attack may have also damaged domestic Israeli cyber networks used by civilians.[60] This shows that a certain degree of civilian collateral damage was permissible even if the attack was focused solely on military objectives.

The Georgia Attack

Compared to Estonia, the Russian attack on Georgia in 2008 presents a slightly more recent example of cyber warfare conducted against a former Soviet state in order to achieve tangible disruption and effects beyond CNE-style espionage. On July 20, prior to the military invasion of Georgia by Russian forces, a large-scale DDOS attack shut down Georgian servers. It is the best example to date of cyber weapons being used as a force multiplier for conventional military operations. As the invasion began, the attacks increased and spread to other targets. This ultimately forced the Georgian government to move critical communication services to commercial U.S. sites as their own services were shut down.[61] The attack was likely organized by the Russian government to support its broader political and military objectives in the crisis but executed by loosely affiliated "independent" hackers that strengthened the government's plausible deniability.[62] Like the Estonian attacks, this attack demonstrated no normative constraint prohibiting targeting civilian resources. However, also like the attacks on Estonia, critical infrastructure was not attacked and permanent damage did not occur. Both of these attacks on former Soviet states—likely originating from the same source—show that the only constraint on the attacks was not a norm; rather it was the limits of what was technologically possible and effective.

The Stuxnet Attack

Perhaps the most famous example of cyber warfare is Stuxnet. In July 2010 a Belarusian computer security firm first identified Stuxnet, an extremely sophisticated computer virus designed to attack industrial control systems.[63] As the global computer security industry began deconstructing the virus, it became apparent that the Iranian nuclear program was its likely target. Soon software patches were posted to eliminate the vulnerabilities Stuxnet exploited, and tools were provided that computer users, including those in Iran, could use to clean their infected machines.[64] The need for these cleanup tools was widespread: in a little over a year, Stuxnet spread prodigiously to approximately 100,000 computers worldwide, 40,000 of which were located outside Iran.[65] Stuxnet's sophistication was in how it spread to a system not connected to the broader Internet, targeted a very specific industrial control system, and fooled operators into thinking everything was normal while wreaking physical havoc on the system.[66] Discounting the 1982 Siberian pipeline attack, Stuxnet was the first incident of cyber warfare that targeted physical infrastructure and caused real-world damage without involving any kinetic weapons. It was, in the words of former CIA director Michael Hayden, "the first attack of a major nature in which a cyber attack was used to effect physical destruction rather than just slow another computer, or hack into it to steal data."[67]

Stuxnet utilized many "zero-day" software strategies, precisely identified its targets, and activated its destructive payload only when it found the specific Siemens programmable logic controller (PLC) used for Iranian centrifuges.[68] Zero-day attacks take advantage of previously unknown vulnerabilities in a computer application. When the target was identified, Stuxnet modified the code on the Siemens PLC in order to cause physical damage while simultaneously masking its modifications to make the system appear to be functioning normally. Experts projected that this likely delayed the Iranian nuclear program by six to eighteen months and destroyed approximately one thousand P-1 centrifuges, 20 percent of Iran's total inventory.[69] Once the public became aware of Stuxnet, there was immediate suspicion that the United States and Israel were behind the attack. Nevertheless, as with the attacks on Estonia and Georgia as well as the Conficker virus, conclusive attribution was not possible. However, in June 2012 a *New York Times* story based on unspecified U.S. sources stated that Stuxnet was part of a series of U.S. cyber attacks organized under the code name Olympic Games.[70] The journalist David Sanger reported that even after Stuxnet became public, the United States allegedly decided to accelerate additional cyber attacks on Iran, perhaps due to the remarkable success of Stuxnet. In October 2013 a cyber researcher named Ralph Langner reported that Stuxnet actually had two attack protocols; the widely reported centrifuge overspinning attack was the simpler and less severe payload. The second Stuxnet payload would have overpressurized Iran's centrifuges by tampering with the protection system, destroying hundreds of centrifuges at once, but it would have blown Stuxnet's cover, which, Langner argues, is the reason it was not deployed.[71] In terms of cyber warfare norms, the Stuxnet attack showed that significant cyber attacks against sensitive and critical targets are acceptable. Iran was reluctant to even acknowledge the attack, perhaps in part because it did not believe the action was prohibited under customary international law.[72] Other nations may also have fallen victim to Stuxnet as collateral damage; in November 2013 Kaspersky claimed that Stuxnet also infected nuclear facilities outside Iran, including a Russian nuclear plant.[73]

The Saudi Aramco Attack

In part as a response to the damage wrought by Stuxnet, Iran is believed to have invested heavily in offensive cyber warfare capabilities. On August 15, 2012, these investments bore fruit in an attack involving the Shamoon virus that was launched against the state-owned oil company Saudi Aramco, the most valuable company in the world.[74] The attack prompted Secretary of Defense Leon Panetta to describe Shamoon as a "very sophisticated" piece of malware generating "tremendous concern."[75] Over thirty thousand computers were infected, and in many cases data on servers as well as hard drives on individual computers were destroyed.[76] The goal of the attack was purportedly to disrupt the flow of Saudi oil by damaging SCADA control systems, but it did not succeed in achieving that effect.[77] An Iranian-linked group called Cutting Sword of Justice ultimately took credit for the attack, which also affected the Qatari company RasGas as well as other oil companies.[78] The attack affected the business processes of Saudi Aramco, and it is likely that some important drilling and production data were lost.[79] Shamoon followed the dangerous trend of unconstrained attacks against nonmilitary targets; Richard Clarke, a cyber warfare expert and former senior official at the U.S. National Security Council, saw it as a signal that this kind of retaliation and escalation was just beginning.[80] This is not a positive sign for those hoping for the emergence of a constraining or limiting norm for cyber warfare.

Operation Ababil

In September 2012, not long after the Saudi Aramco attacks, further retaliation and escalation stemming from the Stuxnet attack on Iran occurred when the Iranian-affiliated hacker group Izz ad-Din al-Qassam launched Operation Ababil, targeting the websites of financial institutions for major DDOS attacks. There institutions included the Bank of America, the New York Stock Exchange, Chase Bank, Capital One, SunTrust, and Regions Bank.[81] In January 2013 Izz ad-Din al-Qassam claimed responsibility for another series of DDOS attacks again predominantly U.S. financial institutions as part of Operation Ababil Phase 2. A third phase of DDOS attacks began in March 2013.[82] U.S. officials believe that Izz ad-Din al-Qassam is a front organization for an Iranian state-sponsored effort.[83] Senator Joseph Lieberman went so far as to state on C-SPAN, "This was done by Iran and the Quds Force, which has its own developing cyberattack capability."[84] Unfortunately, given the at best ambiguous attribution of major cyber attacks, let alone the daily drone of CNE, norms constraining cyber warfare do not appear to be emerging. The current environment allows states to view their own attacks as retaliation and not escalation (as Iran surely does following Stuxnet) and thus feel even less constrained by any sense of appropriate and inappropriate targets and methods. Table 10 summarizes these attacks and their implications for norm emergence.

Examples of Restraint

In addition to actual attacks, decisions *not* to employ cyber warfare also indicate an emerging customary practice and potential nascent norm. Fear of collateral damage of civilian targets and an inability to discriminately target military objectives have led some cyber warfare plans to be called off. For example, in advance of a physical invasion in 2003, the United States was planning a massive CNA-style cyber attack on Iraq to freeze bank accounts and cripple and disrupt government systems. Despite possessing the ability to carry out such attacks, the Bush administration canceled the plan out of a concern that the effects would not be contained to Iraq but instead would also have a negative effect on the civilian networks of allies across the region and in Europe.[85] In 2011 the United States allegedly considered using cyber weapons to disrupt Libya's air defenses but chose not to, in part due to the limited time available and greater suitability of conventional weapons to achieve the desired effects.[86] The United States again declined to launch kinetic or cyber attacks against the Syrian regime in August and September 2013, in part due to concerns about causing unintentional collateral damage as well as concerns regarding Syria's and Iran's ability to retaliate in cyberspace against U.S. banks and other targets (which Iran did following Stuxnet).[87] This restraint offers a glimmer of hope for the emergence of constraining norms for cyber warfare, although other factors and practical considerations likely played a role in these examples of nonuse. Healey identifies this restraint against engaging in "full-scale strategic cyber warfare" as a de facto norm.[88] That said, no state has protested any of these CNA-style cyber attacks as a violation of international law, although Georgia did protest the ongoing physical invasion of their country by Russian forces.[89] This further suggests that this level of offensive cyber warfare is permitted by existing norms.

Current Efforts to Reach Consensus and Codify Norms

While the preceding information makes it apparent that few, if any, normative constraints governing cyber warfare exist, increased attention and discussion have helped spur efforts to reach a consensus on and codify emerging norms for cyber warfare. Norm evolution theory indicates that norm emergence is more likely to occur when norm entrepreneurs with organizational platforms and key states acting as norm leaders are involved. The two primary intergovernmental bodies and organizational platforms currently being used to promote emerging norms for cyber warfare are the UN and NATO. Other key multilateral efforts to encourage the development of cyber norms are the London Conference on Cyberspace (and subsequent conferences) and academic cyber norm workshops. Efforts in the UN have primarily been led by Russia, while efforts in NATO have been led by the United States. In addition to these two main forums focused on cyber warfare, the European Union's Council of Europe Convention on Cybercrime is a regional yet important treaty that went into force in 2004.[90] The convention criminalizes nonstate cyber crime and obliges states parties to prevent nonstate actors from launching cyber attacks from their territory.[91] Additionally the UN has organized a series of events under the umbrella of the World Summit on the Information Society, which, like the European Union's efforts, includes actions against cyber crime.[92] However, these other efforts are only indirectly focused on the issue of CNA-style cyber warfare conducted between nation-states.

The UN as an Organizational Platform for Cyber Norm Emergence

Since the UN Charter entered into force in 1945, international law and norms have been based on Article 2(4), which directs that "all Members shall refrain in their international relations from the threat or use of force against the territorial integrity or political independence of any state, or in any other manner inconsistent with the Purposes of the UN."[93] How does cyber warfare fit into this construct?

Within the UN the main focus on cyber warfare has occurred in the UN General Assembly's First Committee (Disarmament and International Security Committee) as well as various subsidiary organs and specialized agencies, particularly the ITU, UNIDIR, and the CTITF working group.[94] Serious focus on cyber warfare began in 1998 with the Russian resolution in the First Committee, "Developments in the Field of Information and Telecommunications in the Context of Security," to establish cyber arms control similar to other arms control agreements.[95] Richard Clarke, a former senior official who led the U.S. government opposition to the treaty, "viewed the Russian proposal as largely a propaganda tool, as so many of their multilateral arms control initiatives have been for decades."[96] The U.S. position is that the LOAC apply to state behavior in cyberspace and so a prohibition on offensive cyber weapons is unnecessary. Interestingly, China, another key actor in cyberspace, has largely been quiet on the Russian proposals and efforts within the UN to develop and codify cyber norms and has supported them only in recent years.[97] While the Russian proposal in 1998 was adopted by the General Assembly without a vote every year between 1998 and 2004 (a sign of a lack of consensus and weak reception), in 2005 a vote was taken; fourteen other nations, including China, signed on as cosponsors, and the United States was the only country voting against the resolution.[98] The proposal established a group of government experts (GGE) in 2004, which raised the profile of the issue of cyber warfare but failed to achieve consensus on whether the LOAC were sufficient to address the threat. Perhaps due to the cyber attacks in Estonia and Georgia, the Russian proposal reverted back to being adopted without a vote in 2009.[99] Then, in 2010, the United States reversed its opposition and supported adoption, perhaps a tactical move of the Obama administration's "reset" with Russia.[100]

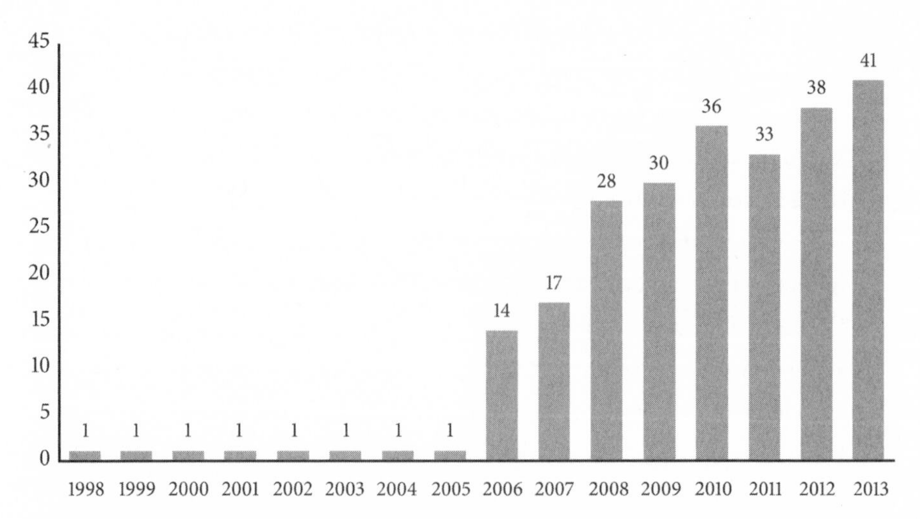

FIG. 1. Number of state cosponsors of UN First Committee cyber resolution (1998–2013). *Source:* Maurer, "Cyber Norm Emergence at the United Nations," 26.

The United States also began cosponsoring the proposal in 2010, along with thirty-five other nations. Clearly momentum is building; figure 1 is based on information compiled by Tim Mauer as well as information available in UN databases.[101]

The draft UN cyber resolution the United States supported in 2010 lacked the important reference to the need to develop definitions of key terms, which would be the first real step in developing a cyber arms control treaty.[102] While momentum in favor of the Russian proposal is clearly building, this change also weakens the tangible effect of the resolution insofar as it could actually lead to a binding agreement for cyber warfare. In addition to this annual cyber resolution, in 2011 Russia and China offered an additional proposal, "International Code of Conduct for Information Security," which has the "aim of achieving the earliest possible consensus on international norms and rules guiding the behavior of States in the information space."[103] This proposed code, which has eleven main points, seems to tack back against U.S. and Western concerns regarding Russian interests in limiting Internet freedom by prohibiting not only CNA-style attacks but also "information warfare" and the free exchange of ideas.[104]

In addition to growing support for the Russian cyber resolution (as amended) in the First Committee, there are other indicators of activity to address cyber warfare in the UN. Unlike the 2004 GGE, which failed to achieve consensus, more recent GGEs have had far greater success. In 2010 and 2013 the GGEs established by the First Committee cyber resolutions were able to achieve consensus and generate several recommendations.[105] For example, the 2010 GGE consensus report called for a sustained dialogue on "norms for state use of information and communications technologies to reduce risk and protect critical infrastructures" and "confidence-building and risk reduction measures, including discussion of information and communication technologies in conflict."[106] The GGE's 2013 report broke new ground in affirming that "the application of norms derived from existing international law" was relevant to cyber warfare and "essential to reduce risks to international peace, security and stability." This seems to represent a break toward the long-standing U.S. position that existing international law and agreements regarding the use of force are sufficient to address the new challenge of cyber warfare. The 2013 GGE report also recommends additional study to promote shared understanding regarding how this existing international law and norms apply to state behavior in cyberspace given the unique characteristics of cyber warfare, noting that "additional norms could be developed over time." This last component could be a nod toward the Russian position that new and more constraining and specific norms (such as an outright ban or prohibition on first use) could eventually be adopted. Finally, the 2013 GGE offered voluntary confidence-building measures (CBMs) to promote trust, increase predictability, and reduce misperception. These include measures such as exchanging views and information on national strategies, policies, and organizations as well as the creation of bilateral and multilateral consultative frameworks such as seminars and workshops.[107] This echoes the conclusions of the Security Defense Agenda's 2012 report, which called for cyber CBMs as an alternative to a global treaty or at least as a near-term stopgap measure.[108] ICT4Peace, an international organization that spun off the UN's World Summit on the Information Society activities, has taken a leadership role in developing cyber CBMs and issued a report in October 2013 identifying a process to do so.[109]

In addition to the efforts of the First Committee, the ITU, UNIDIR, and CTITF working group have also taken steps to promote the emergence of norms for cyber warfare. Of these, the ITU, as a treaty-level UN organization, is perhaps the most significant organizational platform for norm emergence. The ITU, partnering with the World Federation of Scientists, initially approached cyber warfare at the request of various states but has since acted autonomously as a norm entrepreneur in pursuit of its own "cyber peace" agenda seeking to prevent the use of force in cyberspace.[110] The ITU secretary-general submits quarterly cyber threat assessments to the UN secretary-general and maintains a database of experts to be consulted in the event of a major cyber attack.[111] Much of the ITU's cyber efforts are focused more on general cyber hygiene and cyber crime, but it has advocated for norms related to cyber warfare. At the 2010 World Telecom Development Conference, the ITU secretary-general proposed a "no first attack vow" for cyber warfare as well as an obligation of states to prevent independent or nonstate attacks from originating from their territory.[112] With the World Federation of Scientists, the ITU helped develop various declarations, such as the Quest for Cyberpeace, which advocate for these two norms.[113]

UNIDIR has also played a role in fostering cyber norms. Germany has acted as a key norm leader by sponsoring ongoing UNIDIR research titled "Perspectives on Cyber War: Legal Frameworks and Transparency and Confidence Building."[114] Russia previously sponsored much of UNIDIR's cyber warfare–related activities. UNIDIR's effort seeks to raise general awareness of cyber warfare and generate multilateral discourse through publications and various meetings and conferences. UNIDIR staff also serve the GGE sessions supporting the First Committee's efforts. Within the overarching UN framework, the CTITF Working Group on Countering the Use of the Internet for Terrorist Purposes provides an organizational platform for examining cyber warfare issues, albeit from a nonstate actor focus. The group's initial report in February 2009 included the concerns of two states (out of thirty-one participating) regarding nonstate actor CNA-style cyber attacks.[115]

NATO *as an Organizational Platform for Cyber Norm Emergence*

Following the major cyber attacks on Estonia (a NATO member) in 2007 and Georgia (an aspiring NATO member) in 2008, NATO began to focus more seriously on the threat of cyber warfare.[116] In 2008 NATO established the NATO CCD COE. located in Tallinn, Estonia.[117] Its mission is to enhance NATO's cyber defense through research, education, and consulting. In 2012 the organization published *National Cyber Security Framework Manual* to help member nations better develop national policies for cyber defense. NATO's commitment to addressing cyber warfare extends beyond this center of excellence.

In November 2010, NATO adopted a new strategic concept, which recognized that cyber warfare "can reach a threshold that threatens national and Euro-Atlantic prosperity, security and stability."[118] These efforts have in part been motivated by increasing interest and concern from the public regarding the cyber threat. For example, in 2013 the National Geographic channel aired a movie titled *American Blackout*, which depicted chaos and destruction following a massive cyber attack that took out the U.S. power grid.[119] In general, NATO, led by the United States, has approached cyber warfare from a perspective that seeks to apply the existing LOAC to cyber attacks rather than pursue more comprehensive and new restrictions like those proposed by Russia in the UN. NATO's most important activity in this effort was the development of the *Tallinn Manual on the International Law Applicable to Cyber Warfare*. The *Tallinn Manual*, which does not reflect official NATO opinion but rather the personal opinion of the authors (an "international group of experts"), was sponsored by the NATO CCD COE and three organizations acting as observers: NATO, CYBERCOM, and the International Committee of the Red Cross.[120] Also noteworthy is an independent yet similar effort by Israel, led by Col. Sharon Afek, which reached similar conclusions regarding the LOAC and cyber warfare in early 2014.[121] The *Tallinn Manual* represents not only the consensus view of these NATO-affiliated participants, but also the main positions of the U.S. government.[122] This is based on a September 2012 speech by a U.S. State Department legal advisor, Harold Koh, who articulated the U.S. positions on international law and cyberspace, which are consistent with the positions articulated in the *Tallinn Manual*.[123] In addition, the 2011 "International Strategy for Cyberspace" specified that the "long-standing international norms guiding state behavior—in times of peace and conflict—also apply in cyberspace."[124] Both the Koh speech and the *Tallinn Manual* go further to flesh out the U.S.-NATO position that the international LOAC are adequate and applicable to cyber warfare and reject the Russian position that cyber warfare requires new and distinct international norms and agreements. This argument was made in the past by other U.S. leaders, including Gen. Keith Alexander, commander of CYBERCOM, who testified that military operations in cyberspace "must comply with international law that governs military operations," essentially equating cyber weapons with guns and bombs and implying that the rules that apply to those also apply to cyber warfare.[125] The interpretation of the LOAC to cyber warfare can be challenging, and a consensus on application needs to be developed.[126] This lack of clarity is similar to the confusion over definitions and application of norms for airpower in the early part of the twentieth century. The following areas of application are currently being resolved: What constitutes the use of force in cyberspace? What constitutes an armed attack in cyberspace? What constitutes legitimate military objectives in cyberspace, and how does the principle of distinction and noncombatant immunity apply? What role does state sovereignty play in cyberspace?[127] Both the United States (as interpreted through the Koh speech) and the group of experts that developed the *Tallinn Manual* attempt to clarify these issues. For example, they identify cyber attacks that "result in death, injury, or significant destruction" as the use of force in cyberspace, and they determined that "whether a cyber use of force qualifies as an armed attack" depends on its "scale and effects" and that cyber attacks must exercise distinction and be aimed at legitimate military objectives, although defining a military objective in cyberspace is particularly complicated.[128] This last point is an area where the United States may stand apart from the authors of the *Tallinn Manual* in supporting a fairly broad definition of military objectives that includes "war-sustaining" objectives in addition to "war-fighting" objectives.[129]

Media response to the *Tallinn Manual* has varied. Some have misinterpreted and twisted its conclusions to claim that it justifies "killing hackers."[130] However, it represents a significant step forward in developing the emerging cyber warfare norm tied to the existing war-fighting norms codified in the LOAC. Today all major powers except China agree that the LOAC apply to cyber warfare.[131]

Other Multilateral Forums Acting as Organizational Platforms for Cyber Norm Emergence

The United Kingdom has been the most active individual norm entrepreneur. Its Foreign and Commonwealth Office hosted the London Conference on Cyberspace in November 2011 to "discuss the vital issues posed for us all by a networked world connected ever more closely together in cyberspace." The conference involved over seven hundred participants from sixty nations, including Prime Minister David Cameron and Vice President Joseph Biden. Among the many cyber topics addressed, discussions included international security and cyber warfare. The conference chair reported that "all delegates underlined the importance of the principle that governments act proportionately in cyberspace and that states should continue to comply with existing rules of international law and the traditional norms of behavior that govern interstate relations, the use of force and armed conflict, including the settlement by states of their international disputes by peaceful means in such a manner that international peace, security and justice are not endangered." The participants agreed the next steps should be to focus on further developing "shared understanding" (norms) through efforts such as the UN First Committee's

GGE and other organizations. However, the participants did not have the "appetite . . . to expend effort on legally-binding international instruments."[132] Following the conference in London, a follow-up conference was held in Budapest in October 2012, and then in Seoul in October 2013.[133] The conference in Seoul had over a thousand participants from approximately ninety countries and generated the "Seoul Framework for and Commitment to Open and Secure Cyberspace" as well as plans for a fourth conference, in 2015 in The Hague.[134] The "Seoul Framework" reaffirmed the conclusion of the London conference that existing norms and international law apply to cyber warfare and that states should prevent nonstate actors from launching attacks from their territory. The framework also noted that additional norms could be developed over time.[135]

In addition to this U.K.-initiated global multilateral effort to address cyber issues, including cyber warfare, regional multilateral efforts have included efforts at the Organization for Security and Cooperation in Europe and the Association of Southeast Asian Nations Regional Forum.[136] These groups have begun discussions on "cyber confidence-building measures" heading into 2014.[137] CBMs have been used to reduce uncertainty and potential for miscalculation with other weapons. As part of the 1986 Second Review Conference of the BWC, states parties agreed to implement a number of CBMs in order to "prevent or reduce the occurrence of ambiguities, doubts and suspicions and in order to improve international co-operation in the field of peaceful biological activities."[138] Another example of a regional multilateral effort to cultivate norms for cyber warfare is a group of government leaders and security experts (eighty senior military officials from twenty-three Asia-Pacific nations), met in Seoul less than a month after the 2013 Seoul Cybersecurity Conference for the Seoul Defense Dialogue, which also addressed cyber warfare.[139] Chinese professor Jia Qingguo summarized the results of the defense dialogue, saying that clearer definitions of cyber warfare and rules of engagement were needed and that "those who engage in cyber warfare should make necessary efforts to avoid attacking civilian infrastructures that may harm the civilians . . . [and] major countries should work together to develop an agreement on a code of conduct in cyberspace and a set standards."[140] There are also numerous bilateral dialogues that involve key cyber actors, such as China, the United States, and Russia.[141] This includes activities such as the cyber agreement Russia and the United States signed in June 2013, which established a communication hotline for a cyber crisis. In light of North Korea's increasingly bellicose cyber posture, South Korea and the Unites States established a Cyber Cooperation Working Group in early 2014 to discuss cooperation on cyber warfare issues.[142]

Other Academic Forums Acting as Organizational Platforms for Cyber Norm Emergence

Various academic and private entities have served as organizational platforms and norm entrepreneurs. Massachusetts Institute of Technology (MIT) held a "Cyber Norms Workshop" in October 2011 in Cambridge, Massachusetts, sponsored by the University of Toronto's Canada Center for Global Security Studies, the Belfer Center for Science and International Affairs at Harvard's Kennedy School of Government, and MIT and Microsoft's Explorations in International Cyber Relations. The workshop concluded that developing a broad set of comprehensive norms for cyber warfare may be too difficult; instead different norms may develop at various levels (within groups of states, bilaterally, and so on). Consistent with recent parallel efforts in the UN and NATO, the participants also agreed that key existing international norms, including the LOAC, should be applied to cyber warfare and that cyber CBMs are needed.[143] The participants noted that while the LOAC should apply as a norm in cyberspace, application was very complex, and the principles of noncombatant immunity and proportionality would be difficult to operationalize.[144] A second workshop was held in Cambridge in September 2012 and further developed these thoughts as well as some new concepts; for example, the participants contemplated a cyber risk reduction scheme based on a regime enforced by major cyber players, similar to the nuclear regime imposed by the Nuclear Non-Proliferation Treaty when it codified the legitimacy of the existing nuclear powers.[145] However, such an effort may be fruitless as it is predicated on the assumption that states can be prevented from developing cyber weapons, a nearly impossible task.

Additional academic efforts to examine norms for cyberspace and cyber warfare are under way with government support. The National Science Foundation recently awarded two $500,000 grants to professors at Western Michigan University and the Naval Postgraduate School to study the ethics of cyber warfare.[146] Think tanks have helped promote dialogue on norms for cyber warfare. The EastWest Institute (EWI) has organized consultations in Washington, Moscow, and Brussels to discuss cyber diplomacy, resulting in a report outlining how the two key states, Russia and the United States, could work together to find common ground on norms for cyber warfare (among other things), such as conducting joint policy assessments of international cyber law. This effort hopes to capitalize on the "thaw" in cyber diplomacy in 2010 after the United States joined Russia in supporting a modified version of its cyber resolution in the UN First Committee.[147] EWI also organized a series of Worldwide Cybersecurity Summits beginning in 2010 and published a manual identifying common cyber terminology in Russia and the United States.[148] Other academic organizations, such as the NRC—part of the U.S. National Academies—have developed concepts and norms associated with cyber warfare. In 2009 the NRC published a groundbreaking book arguing that cyber warfare was subject to the LOAC and UN Charter. The authors noted that, because of the novel nature of cyber weapons, there were "uncertainties in how LOAC and UN Charter law might apply in any given instances" and that application becomes particularly complex when an attack's "effects do not entail physical damage or loss of life but do have other negative effects on another nation."[149] These academic activities and venues have helped prompt norm emergence by serving as organizational platforms (often with a domestic or bilateral focus) for norm entrepreneurs and leaders.

NOTES

INTRODUCTION

1. Hugo, *The History of a Crime*, 364.
2. Katzenstein, Wendt, and Jepperson, "Norms, Identity, and Culture," 54.
3. Gavin, *War and Peace in the Space Age*, 265.
4. Thomas, *The Ethics of Destruction*, 97–98.
5. James R. Clapper, "Statement for the Record: Worldwide Threat Assessment of the U.S. Intelligence Community," Senate Select Committee on Intelligence, March 12, 2013, 1–3, http://www.intelligence.senate .gov/130312/clapper.pdf.
6. *International Strategy for Cyberspace: Prosperity, Security, and Openness in a Networked World*, WhiteHouse.gov, May 2011, http://www .whitehouse.gov/sites/default/files/rss_viewer/international_strategy _for_cyberspace.pdf.
7. Eric Chabrow, "Using Diplomacy to Stop Cyber-Attack," GovInfoSecurity.com, March 1, 2013, http://www.bankinfosecurity.eu /using-diplomacy-to-stop-cyber-attacks-a-5572; Mark Landler and David Sanger, "U.S. Demands China Block Cyberattacks," *New York Times*, March 11, 2013, http://www.nytimes.com/2013/03/12/world/asia /us-demands-that-china-end-hacking-and-set-cyber-rules.html?_r=0.
8. Katzenstein, Wendt, and Jepperson, "Norms, Identity, and Culture," 54.
9. For the purposes of this book, WMD include nuclear, biological, and chemical weapons. This study reviews the three WMD variants in discrete

sections (within two chapters) as there are important and significant differences between norms on possession and use for nuclear, chemical, and biological weapons.

10. Brian Fung and Michelle Boorstein, "Pope Francis Calls the Internet 'A Gift from God,'" *Washington Post*, January 23, 2014, http://www.washingtonpost.com/blogs/the-switch/wp/2014/01/23/the-pope-calls-the-internet-a-gift-from-god/.
11. U.S. Department of Defense, *Joint Publication 1-02*, 93.
12. Dennis Murphy, "What Is War? The Utility of Cyberspace Operations in the Contemporary Operational Environment," issue paper vol. 1-10, Center for Strategic Leadership, U.S. Army War College, February 2010, http://www.dtic.mil/dtic/tr/fulltext/u2/a518178.pdf.
13. Campen, "Rush to Information-Based Warfare Gambles with National Security."
14. U.S. Government Accountability Office, *Defense Department Cyber Efforts*, 10.
15. U.S. Department of Defense, *Joint Publication 1-02*, 93.
16. Brainy Quote http://www.brainyquote.com/quotes/quotes/y/yogiberra141506.html.
17. While chemical and biological weapons, strategic bombing, and nuclear weapons are very distinct issues, the respective development of international regulative norms for each weapon type while they were emerging-technology weapons may be illustrative regarding cyber norms. These three case studies were selected using John Stuart Mill's method of agreement, whereby cases are selected that take on the same values of the dependent variable (cyber warfare), as indicated earlier. However, there are some independent variables that vary across case studies (that is, the availability of nuclear weapons and bombers is vastly different) that should help increase the robustness of the methodology.
18. Koblentz and Mazanec, "Viral Warfare."
19. Overy, *The Air War*, 24.
20. Thomas Schelling, "The Nuclear Taboo," *MIT International Review* (Spring 2007), http://web.mit.edu/mitir/2007/spring/taboo.pdf.

1. NORM EVOLUTION THEORY

Epigraph: Victor Hugo, *The History of a Crime* (Whitefish MT: Kessinger, 2004).

1. United States, "International Strategy for Cyberspace: Prosperity, Security, and Openness in a Networked World," May 2011, http://www.whitehouse.gov/sites/default/files/rss_viewer/international_strategy_for_cyberspace.pdf.
2. Katzenstein, Wendt, and Jepperson, "Norms, Identity, and Culture," 54.
3. Florini, "Evolution of International Norms," 364.
4. Finnemore and Sikkink, "International Norm Dynamics," 887.
5. DeLamater, "Social Control of Sexuality."
6. "Saudi Minister Rebukes Religious Police," BBC News, November 4, 2002, http://news.bbc.co.uk/2/hi/middle_east/2399885.stm.
7. Katzenstein, Wendt, and Jepperson, "Norms, Identity, and Culture," 33–75.
8. Gavin, *War and Peace in the Space Age.*
9. Thomas Schelling, "The Nuclear Taboo," *MIT International Review* (Spring 2007), http://web.mit.edu/mitir/2007/spring/taboo.pdf.
10. Paul, *The Tradition of Non-Use of Nuclear Weapons.*
11. Finnemore and Sikkink, "International Norm Dynamics," 891.
12. Legro, "Which Norms Matter?," 32.
13. Genest, *Conflict and Cooperation*, 42.
14. Tarzi, "The Role of Principles, Norms and Regimes," 9.
15. Tarzi, "The Role of Principles, Norms and Regimes," 11.
16. Rose, "Neoclassical Realism," 144–72.
17. Sagan, *Ethics and Weapons of Mass Destruction*, 73–95.
18. Puchala and Fagan, "International Politics in the 1970s," 247–66.
19. Keohane and Nye, *Power and Interdependence*, 24–25.
20. Nye and Keohane, *Governance in a Globalizing World*, 1.
21. Wendt, *Social Theory of International Politics.*
22. Bull, *The Anarchical Society*, 2.
23. Tarzi, "The Role of Principles, Norms and Regimes," 10.
24. Tarzi, "The Role of Principles, Norms and Regimes," 10.

25. Glenn, "Realism versus Strategic Culture," 523–51.
26. For example, see Goertz, *International Norms and Decision Making*; Rublee, *Nonproliferation Norms*; Finnemore and Sikkink, "International Norm Dynamics"; Checkel, "Institutions, and National Identity," 83–114; Katzenstein, Wendt, and Jepperson, "Norms, Identity, and Culture"; Ratner, "International Law," 65–80.
27. Florini, "The Evolution of International Norms," 363–89.
28. Legro, "Military Culture and Inadvertent Escalation," 108–42; Stephen Wrage, "Compliance with Aerial Bombing Norms: A Study of Two Periods, 1939–1945 and 1990–2004," Annual Convention of the Joint Services Conference on Professional Ethics (2004), http://isme.tamu.edu/JSCOPE04/Wrage04.html.
29. University of California, Berkley, Understanding Evolution website, http://evolution.berkeley.edu/evolibrary/article/evo_25.
30. Florini, "The Evolution of International Norms," 364.
31. A phenotype is an organism's observable characteristics or traits.
32. Florini, "The Evolution of International Norms," 367–68.
33. Finnemore and Sikkink, "International Norm Dynamics," 887–917.
34. Florini, "The Evolution of International Norms," 367–68.
35. Garcia, *Small Arms and Security.*
36. Garcia, *Small Arms and Security*, 374.
37. Finnemore and Sikkink, "International Norm Dynamics," 898–917.
38. Schelling, "The Nuclear Taboo."
39. William Potter, "In Search of the Nuclear Taboo: Past, Present, and Future," Institut Français des Relations Internationales, *Proliferation Papers*, no. 31 (Winter 2010), http://cns.miis.edu/other/potter_william_100115_IFRI_pp31.pdf.
40. Finnemore and Sikkink, "International Norm Dynamics," 899.
41. Parlow, "Banning Land Mines," 715–39; Matthew and Rutherford, "The Evolutionary Dynamics of the Movement to Ban Landmines," 29–56.
42. Moravcsik, "The Origins of Human Rights Regimes," 217–52.
43. Tannenwald, *The Nuclear Taboo.*

44. Finnemore and Sikkink, "International Norm Dynamics," 895–901.
45. Finnemore and Sikkink, "International Norm Dynamics,"908.
46. Florini, "The Evolution of International Norms," 376.
47. Acharya, "How Ideas Spread," 239–75.
48. Parlow, "Banning Land Mines," 715–39.
49. For example, see Kartchner, "Weapons of Mass Destruction."
50. Florini, "The Evolution of International Norms," 367–68.
51. Merriam-Webster, "Meme," http://www.merriam-webster.com/dictionary/meme.
52. Price, "Reversing the Gun Sights," 613–44.
53. Duncan MacLeod, "Landmines on the Soccer Field," The Inspiration Room, March 23, 2005, http://theinspirationroom.com/daily/2005/landmines-clearance/.
54. Price, "Reversing the Gun Sights," 613–44.
55. Florini, "The Evolution of International Norms," 377.
56. Florini, "The Evolution of International Norms," 377.
57. O'Dwyer, "First Landmines, Now Small Arms?," 77–97.
58. Florini, "The Evolution of International Norms," 377–79.
59. Keegan, *The Second World War.*
60. Finnemore and Sikkink, "International Norm Dynamics," 901.
61. Finnemore and Sikkink, "International Norm Dynamics," 901, 902–3, 906.
62. Price, *The Chemical Weapons Taboo*; Price, "A Genealogy of the Chemical Weapons Taboo," 73–103.
63. Tannenwald and Price, "Norms and Deterrence," 114–52.
64. Finnemore and Sikkink, "International Norm Dynamics," 909.
65. Finnemore and Sikkink, "International Norm Dynamics," 909.
66. Finnemore and Sikkink, "International Norm Dynamics," 909.
67. Finnemore and Sikkink, "International Norm Dynamics," 904.
68. Finnemore and Sikkink, "International Norm Dynamics," 904.
69. Moravcsik, "The Origins of Human Rights Regimes," 217–52.
70. Finnemore and Sikkink, "International Norm Dynamics," 893.
71. Tannenwald, *The Nuclear Taboo.*

72. Finnemore and Sikkink, "International Norm Dynamics," 905.
73. Finnemore and Sikkink, "International Norm Dynamics," 905.
74. Sagan, "Why Do States Build Nuclear Weapons?," 74.
75. Finnemore and Sikkink, "International Norm Dynamics," 907.
76. Thomas, *The Ethics of Destruction*, 97–98; Bialer, *The Shadow of the Bomber.*
77. Thomas, *The Ethics of Destruction*, 99.
78. Keegan, *The Second World War.*
79. Cole, "The Poison Weapons Taboo," 119–32.
80. Douglass, *Purity and Danger.*
81. For example, see Mona Lena Krook and Jacqui True, "Rethinking the Life Cycles of International Norms: The United Nations and the Global Promotion of Gender Equality," *European Journal of International Relations* (2010), http://ejt.sagepub.com/content/early/2010/11/04/1354066110380963.full.pdf; Susanne Alldén, "How Do International Norms Travel? Women's Political Rights in Cambodia and Timor-Leste," *Department of Political Science, Umea University* (2009), http://www.diva-portal.org/smash/get/diva2:274281/FULLTEXT01; Verdirame, "Testing the Effectiveness of International Norms," 733–68.
82. Legro, "Which Norms Matter?," 31–63.

2. CHEMICAL AND BIOLOGICAL WEAPONS

Epigraph: Glenn Kessler, "President Obama and the 'Red Line' on Syria's Chemical Weapons," *Washington Post*, September 6, 2013, http://www.washingtonpost.com/blogs/fact-checker/wp/2013/09/06/president-obama-and-the-red-line-on-syrias-chemical-weapons/.

1. Koblentz, *Living Weapons*, 9.
2. Organization for the Prohibition of Chemical Weapons, "Brief Description of Chemical Weapons," http://www.opcw.org/about-chemical-weapons/what-is-a-chemical-weapon/.
3. United Nations Office for Disarmament Affairs, Protocol for the Prohibition of the Use in War of Asphyxiating, Poisonous or Other Gases, and of Bacteriological Methods of Warfare, http://disarmament.un.org/treaties/t/1925.

4. The Biological and Toxin Weapons Convention website, Department of Peace Studies of the University of Bradford, List of States Parties to the Convention on the Prohibition of the Development, Production and Stockpiling of Bacteriological (Biological) and Toxin Weapons and on Their Destruction as of June 2005, https://www.opcw.org/index.php?eID=dam_frontend_push&docID=3030.
5. Cole, "The Poison Weapons Taboo," 119.
6. Cole, "The Poison Weapons Taboo," 119; Shoham, "Chemical and Biological Weapons in Egypt," 48–49.
7. Price, *The Chemical Weapons Taboo*, 20–21.
8. Sociobiology is a scientific field based on the assumption or concept that social behavior is a result of biological evolution and attempts to explain and examine social behavior within that context. For more, see Alcock, *The Triumph of Sociobiology.*
9. Mandelbaum, *The Nuclear Revolution*, 38–39.
10. Cole, "The Poison Weapons Taboo," 120.
11. Price, "A Genealogy of the Chemical Weapons Taboo," 80.
12. Cole, "The Poison Weapons Taboo," 130.
13. Cole, "The Poison Weapons Taboo,"120.
14. Mark Wheelis, "Biological Warfare at the 1346 Siege of Caffa," *Journal of Emerging Infectious Diseases* 8, no. 9 (2002), http://wwwnc.cdc.gov/eid/article/8/9/01-0536_article.
15. Cole, "The Poison Weapons Taboo," 120, 124.
16. Wansbrough, *The New Jerusalem Bible.*
17. Price, "A Genealogy of the Chemical Weapons Taboo," 80; Grotius, *The Law of War and Peace*, 3:15–16.
18. Grotius, *The Law of War and Peace*, 3:4.
19. Cole, "The Poison Weapons Taboo," 120.
20. Cultural Policy Research Institute, 1874 Brussels Declaration, http://www.cprinst.org/Home/cultural-property-laws/1874-brussels-declaration.
21. Hudson, "Present Status of the Hague Conventions," 114–17.
22. Katzenstein, Wendt, and Jepperson, "Norms, Identity, and Culture," 127.

23. Katzenstein, Wendt, and Jepperson, "Norms, Identity, and Culture," 131. Previously, similar dehumanizing "civilization" distinctions allowed British general Sir Jeffrey Amherst to use smallpox-infected blankets to attack Indians in North America without violating the ancient CBW norm.
24. Price, "A Genealogy of the Chemical Weapons Taboo," 83.
25. Price, "A Genealogy of the Chemical Weapons Taboo," 83.
26. "Germans Introduce Poison Gas," History Channel, http://www.history.com/this-day-in-history/germans-introduce-poison-gas.
27. Katzenstein, Wendt, and Jepperson, "Norms, Identity, and Culture," 129.
28. Cole, "The Poison Weapons Taboo," 120.
29. Price, "A Genealogy of the Chemical Weapons Taboo," 90, 91.
30. "Washington Naval Conference," United States History, http://www.u-s-history.com/pages/h1354.html.
31. Katzenstein, Wendt, and Jepperson, "Norms, Identity, and Culture," 128.
32. Price, "A Genealogy of the Chemical Weapons Taboo," 92.
33. U.S. Department of State, *Conference on the Limitation of Armament*, 730.
34. Katzenstein, Wendt, and Jepperson, "Norms, Identity, and Culture," 128.
35. U.S. Department of State, Bureau of International Security and Nonproliferation, Protocol for the Prohibition of the Use in War of Asphyxiating, Poisonous or Other Gases, and of Bacteriological Methods of Warfare (Geneva Protocol), http://www.state.gov/t/isn/4784.htm.
36. United Nations Office for Disarmament Affairs, Protocol for the Prohibition of the Use in War of Asphyxiating, Poisonous or Other Gases, and of Bacteriological Methods of Warfare; emphasis added.
37. Croddy, *Weapons of Mass Destruction*, 142.
38. Tucker, *War of Nerves*, 1.
39. Cole, "The Poison Weapons Taboo," 120. A point of contention delaying U.S. ratification was differing interpretations regarding how the protocol applied to tear gas and other harassing agents.
40. Katzenstein, Wendt, and Jepperson, "Norms, Identity, and Culture," 127.

41. U.S. Department of State, Geneva Protocol.
42. "Report Alleges Chemical Arms Atrocities by Japan in WWII," *Los Angeles Times*, September 18, 1988, http://articles.latimes.com/1988-09-18/news/mn-3410_1_chemical-weapons.
43. Price, "A Genealogy of the Chemical Weapons Taboo," 74.
44. Price, "A Genealogy of the Chemical Weapons Taboo," 75; Brown, *Chemical Warfare*, 290.
45. Harris and Paxman, *A Higher Form of Killing*, 135; Ochsner, *The History of German Chemical Warfare*, 23. In addition to international norms, other factors impacted the German's decision, such as the fear of a massive retaliation in kind and the logistical complication offered by Hermann Goering following the war that the Wehrmacht needed horse-drawn transport to resupply their forces and were unable to produce an effective gas mask that would have protected their horses.
46. Price, "A Genealogy of the Chemical Weapons Taboo," 76–77.
47. Price, "A Genealogy of the Chemical Weapons Taboo," 78.
48. Finnemore and Sikkink, "International Norm Dynamics," 904.
49. Finnemore and Sikkink, "International Norm Dynamics," 904.
50. Mandelbaum, *The Nuclear Revolution*, 37.
51. Jonathan Tucker and Erin Mahan, "President Nixon's Decision to Renounce the U.S. Offensive Biological Weapons Program" (Washington DC: Center for the Study of Weapons of Mass Destruction, National Defense University Press, October 2009, http://ndupress.ndu.edu/Portals/68/Documents/casestudies/CSWMD_CaseStudy-1.pdf.
52. Tucker, "A Farewell to Germs," 107–48; Swyter, "Political Considerations and Analysis," 261–70.
53. Dando, *Preventing Biological Warfare*, 5.
54. Jeanne Guillemin, "Scientists and the History of Biological Weapons: A Brief Historical Overview of the Development of Biological Weapons in the Twentieth Century," *Science and Society* 7 (July 2006): 45–49, http://www.ncbi.nlm.nih.gov/pmc/articles/PMC1490304/.
55. Tucker and Mahan, "President Nixon's Decision," 17.

56. United Nations, The Biological Weapons Convention, http://www.unog.ch/80256EE600585943/%28httpPages%29/04FBBDD6315AC720C1257180004B1B2F?OpenDocument.
57. "Text of the Convention on the Prohibition of the Development, Production, and Stockpiling of Bacteriological (Biological) and Toxin Weapons and on Their Destruction," Federation of American Scientists, http://www.fas.org/nuke/control/bwc/text/bwc.htm.
58. Tucker and Mahan, "President Nixon's Decision," 17.
59. Bailey, "Why the United States Rejected the Protocol," 5.
60. Gillian R. Woollett, "Industry's Role, Concerns, and Interests in the Negotiations of a BWC Compliance Protocol," *Stimson Report* 24 (January 12, 1998), http://www.stimson.org/images/uploads/research-pdfs/report24-woollett.pdf.
61. Jonathan Tucker, "Verifying the Chemical Weapons Ban: Missing Elements," *Arms Control Today* (January/February 2007), http://www.armscontrol.org/act/2007_01-02/Tucker.
62. Koblentz, *Living Weapons*, 53–105.
63. Jonathan Tucker, "The Biological Weapons Convention (BWC) Compliance Protocol," *Center for Nonproliferation Studies* (September 2002), http://www.nti.org/e_research/e3_2a.html.
64. Donald A. Mahley, "A Personal Assessment of the BWC Protocol Negotiations," *CBW Conventions Bulletin* (February 2010): 1, http://www.sussex.ac.uk/Units/spru/hsp/documents/CBWCB86.pdf.
65. Koblentz, *Living Weapons*, 143.
66. Arms Control Association, "The Biological Weapons Convention (BWC) at a Glance," http://www.armscontrol.org/factsheets/bwc.
67. Mahley, "A Personal Assessment of the BWC Protocol Negotiations," 2.
68. Seth Brugger, "Executive Summary of the Chairman's Text," Arms Control Association, May 2001, http://www.armscontrol.org/act/2001_05/brugger.
69. Merle D. Kellerhals Jr., "Proposed Biological Weapons Protocol Unfixable, U.S. Official Says," U.S. Department of State, July 25, 2001, http://

sks.sirs.bdt.orc.scoolaid.net/cgi-bin/hst-article-display?id=SNY5703-0-586&artno=0000139161&type=ART&shfilter=U&key=Biological%20Weapons%20Convention%20(1972)&title=Proposed%20Biological%20Weapons%20Protocol%20Unfixable%2C%20U.S.%20Official%20Says&res=Y&ren=Y&gov=Y&lnk=Y&ic=N.

70. Acronym Institute, "Chemical Weapons Convention," http://www.acronym.org.uk/cwc/index.htm.
71. Organization for the Prohibition of Chemical Weapons, "Basic Facts on Chemical Disarmament," http://www.opcw.org/news-publications/publications/history-of-the-chemical-weapons-convention/#c4141.
72. Organization for the Prohibition of Chemical Weapons Technical Secretariat, "Note by the Technical Secretariat."
73. Organization for the Prohibition of Chemical Weapons, "Our Work: Non-proliferation," http://www.opcw.org/our-work/non-proliferation/.
74. Organization for the Prohibition of Chemical Weapons Technical Secretariat, "Status of Participation."
75. Carpenter and Moodie, "Industry and Arms Control," 180.
76. Carpenter and Moodie, "Industry and Arms Control," 178–79.
77. International Council of Chemical Associations, "Statement on the 10th Anniversary of the Chemical Weapons Convention."
78. Organization for the Prohibition of Chemical Weapons, "Chemical Industry and CWC States Parties Meet: Industry Aspects of the Chemical Weapons Ban Discussed," June 13, 2007, http://www.opcw.org/news/article/chemical-industry-and-cwc-states-parties-meet-industry-aspects-of-the-chemical-weapons-ban-discusse/.
79. Hiro, *The Longest War*, 205.
80. Wright, *Dreams and Shadows*, 438.
81. United Nations Security Council, "Report of the Mission Dispatched by the Secretary-General."
82. Patrick Tyler, "Officers Say US Aided Iraq in War Despite Use of Gas," *New York Times*, August 18, 2002, http://www.nytimes.com

/2002/08/18/world/officers-say-us-aided-iraq-in-war-despite-use-of-gas.html?pagewanted=all&src=pm.

83. Katzenstein, Wendt, and Jepperson, "Norms, Identity, and Culture," 129–30.
84. U.S. Department of State, Geneva Protocol.
85. Cole, "The Poison Weapons Taboo," 119.
86. U.S. Central Intelligence Agency, "Intelligence Update: Chemical Warfare Agent Issues," April 2002, https://www.cia.gov/library/reports/general-reports-1/gulfwar/cwagents/index.htm#almuthanna1.
87. Melissa Healey, "Chemical Attack Would Escalate Allied Retaliation," *Los Angeles Times*, February 21, 1991, http://articles.latimes.com/1991-02-21/news/mn-2193_1_chemical-weapons.
88. Cole, "The Poison Weapons Taboo," 129.
89. Kelsey Davenport, "Chronology of Libya's Disarmament and Relations with the United States," Arms Control Association, September 2013, http://www.armscontrol.org/factsheets/LibyaChronology.
90. "Syria Chemical Attack: What We Know," BBC News, September 24, 2013, http://www.bbc.co.uk/news/world-middle-east-23927399.
91. Tim Cohen, "Obama: It's the World's 'Red Line' on Syria. Senate Panel Backs Military Strike Plan," CNN, September 4, 2013, http://www.cnn.com/2013/09/04/politics/us-syria/index.html.
92. Nick Cumming-Bruce, "Watchdog Says Syria Has Been Cooperative on Weapons," *New York Times*, October 9, 2013, http://www.nytimes.com/2013/10/10/world/middleeast/syria-chemical-weapons.html?_r=0; Anne Gearan and Craig Whitlock, "Assad Regime Is Dragging Its Feet on Removing Chemical Weapons, U.S. Says," *Washington Post*, January 30, 2014, http://www.washingtonpost.com/world/national-security/assad-regime-dragging-its-feet-on-removing-chemical-weapons-us-says/2014/01/30/19def3ba-89c7-11e3-916e-e01534b1e132_story.html.
93. Price, "A Genealogy of the Chemical Weapons Taboo," 90.
94. Katzenstein, Wendt, and Jepperson, "Norms, Identity, and Culture," 144.
95. Price, "A Genealogy of the Chemical Weapons Taboo," 98, 78.

96. Price, *The Chemical Weapons Taboo.*
97. Cole, "The Poison Weapons Taboo," 121.
98. Price, "A Genealogy of the Chemical Weapons Taboo," 84.
99. Price, "A Genealogy of the Chemical Weapons Taboo,"94.

3. STRATEGIC BOMBING

Epigraph: Adolf Hitler, speech before the Reichstag, September 1, 1939, http://avalon.law.yale.edu/imt/2322-ps.asp.

1. Longmate, *The Bombers*, 12; Brower, *World War II in Europe*, 108.
2. Kennett, *A History of Strategic Bombing*, 178.
3. Lee, "The Interim Years of Cyberspace," 60.
4. Kennett, *A History of Strategic Bombing*, 1–2.
5. Kennett, *A History of Strategic Bombing*, 2.
6. Kennett, *A History of Strategic Bombing*, 4–5; Davis Biddle, *Rhetoric and Reality*, 19.
7. Davis Biddle, *Rhetoric and Reality*, 12.
8. Pickering, "The Future of Artificial Flight," 17.
9. Davis Biddle, *Rhetoric and Reality*, 12–13.
10. Thomas, *The Ethics of Destruction*, 91.
11. Bellamy, *Just Wars*, 40.
12. U.S. War Department, *The War of the Rebellion*, 148–64.
13. Davis Biddle, "Air Power." 141.
14. Thomas, *The Ethics of Destruction*, 99.
15. Thomas, *The Ethics of Destruction*, 100.
16. Davis Biddle, "Air Power," 141.
17. Thomas, *The Ethics of Destruction*, 100.
18. Davis Biddle, "Air Power," 141.
19. Davis Biddle, "Air Power," 142.
20. Davis Biddle, "Air Power," 143.
21. De Syon, *Zeppelin!*, 25.
22. Davis Biddle, "Air Power," 142.
23. Thomas, *The Ethics of Destruction*, 100.
24. Davis Biddle, "Air Power," 142.

25. Thomas, *The Ethics of Destruction*, 101.
26. Kennett, *A History of Strategic Bombing*, 179.
27. Davis Biddle, "Air Power," 142.
28. Davis Biddle, "Air Power," 143.
29. Adam Roberts, "Undefended Towns," Crimes of War A–Z Law Guide, http://www.crimesofwar.org/a-z-guide/undefended-towns/.
30. Scott, *The Proceedings of the Hague Peace Conference*, 424.
31. Cole and Cheesman, *The Air Defence of Britain*, 6.
32. Quoted in Davis Biddle, "Air Power," 143.
33. H. G. Wells, *The War in the Air*, 1997, BompaCrazy.com, 92–93.
34. Buckley, *Air Power*, 29.
35. Kennett, *A History of Strategic Bombing*, 59.
36. Kennett, *A History of Strategic Bombing*, 59.
37. Kennett, *A History of Strategic Bombing*, 60.
38. Davis Biddle, "Air Power," 144; Thomas, *The Ethics of Destruction*, 103.
39. Davis Biddle, *Rhetoric and Reality*, 21.
40. Davis Biddle, "Air Power," 144.
41. Spaight, *Air Power*, 220.
42. "Gotha G.V. Heavy Bomber (1917)," Military Factory, August 7, 2012, http://www.militaryfactory.com/aircraft/detail.asp?aircraft_id=826.
43. Wise, *Canadian Airmen*, 284–327.
44. Davis Biddle, *Rhetoric and Reality*, 30.
45. Jones, *The War in the Air*, 136.
46. Overy, *The Air War*, 13.
47. Kennett, *A History of Strategic Bombing*, 60–61.
48. Kennett, *A History of Strategic Bombing*, 60.
49. Quoted in Clarke, *Voices Prophesying War*, 100–101.
50. Davis Biddle, *Rhetoric and Reality*, 24.
51. Overy, *The Air War*, 12.
52. Davis Biddle, *Rhetoric and Reality*, 24–25.
53. Davis Biddle, "Air Power," 145.
54. Quoted in Kennett, *A History of Strategic Bombing*, 54.
55. Douhet, *The Command of the Air*, 61.

56. Davis Biddle, *Rhetoric and Reality*, 13.
57. Kennett, *A History of Strategic Bombing*, 54; Hurley, *Billy Mitchell*, 109.
58. Overy, *The Air War*, 13.
59. Lorell, *The U.S. Combat Aircraft Industry*, 38.
60. Thomas, *The Ethics of Destruction*, 118.
61. "Washington Naval Conference," United States History, http://www.u-s-history.com/pages/h1354.html.
62. Davis Biddle, "Air Power," 148.
63. Davis Biddle, "Air Power," 148.
64. Kennett, *A History of Strategic Bombing*, 64.
65. Schindler and Toman, *The Laws of Armed Conflict*, 210–11.
66. Davis Biddle, "Air Power," 148.
67. Quoted in Kennett, *A History of Strategic Bombing*, 65.
68. Kennett, *A History of Strategic Bombing*, 67.
69. Kennett, *A History of Strategic Bombing*, 67.
70. Thomas, *The Ethics of Destruction*, 115.
71. Kennett, *A History of Strategic Bombing*, 69.
72. Thomas, *The Ethics of Destruction*, 116.
73. Germany's withdrawal was not specific to airpower; rather it was a reflection of the new chancellor's (Hitler) nationalistic policy that Germany would no longer accept any reductions in its armed forces and his view that Germany was already disarmed. The Treaty of Versailles prohibited Germany from having an air force; however, Hitler had reestablished the Luftwaffe without consequence.
74. Overy, *The Air War*, 24.
75. Thomas, *The Ethics of Destruction*, 121.
76. Overy, *The Air War*, 104.
77. Overy, *The Air War*, 14.
78. Renz and Scheipers, "Discrimination in Aerial Bombing," 22.
79. Davis Biddle, *Rhetoric and Reality*, 8.
80. Davis Biddle, *Rhetoric and Reality*, 116–17.
81. "The Legacy of Guernica," BBC News, April 26, 2007, http://news.bbc.co.uk/2/hi/europe/6583639.stm.

82. Quoted in Davis Biddle, *Rhetoric and Reality*, 153.
83. Quoted in Parks, "Air War and the Law of War," 39.
84. Murray, "A Tale of Two Doctrines," 86.
85. Davis Biddle, "Air Power," 150.
86. Overy, *The Air War*, 24.
87. Thomas, *The Ethics of Destruction*, 90.
88. Davis Biddle, "Air Power," 150.
89. Spaight, *Air Power*, 43, 259.
90. Thomas, *The Ethics of Destruction*, 91.
91. Davis Biddle, "Air Power," 150.
92. Kennett, *A History of Strategic Bombing*, 112.
93. Legro, "Military Culture," 127.
94. Davis Biddle, "Air Power," 151; Legro, "Military Culture," 128.
95. Webster and Frankland, *The Strategic Air Offensive against Germany*, 205–13.
96. Quoted in Overy, *The Air War*, 115.
97. Quoted in Davis Biddle, *Rhetoric and Reality*, 1.
98. Webster and Frankland, *The Strategic Air Offensive against Germany*, 143–45.
99. Davis Biddle, "Air Power," 152.
100. Overy, *The Air War*, 109.
101. Renz and Scheipers, "Discrimination in Aerial Bombing," 17–18.
102. Davis Biddle, "Air Power," 152.
103. The Norden bombsight was an improved bombsight developed by the United States during World War II. It had an integrated analog computer that was linked to the aircraft's autopilot, which enabled it to achieve dramatic improvements in accuracy. For the radar bombing decision, see Davis Biddle, *Rhetoric and Reality*.
104. Davis Biddle, *Rhetoric and Reality*, 154.
105. Spitzer, "A Forgotten Horror."
106. Tannenwald, *The Nuclear Taboo*, 79.
107. Quoted in Norris, *Racing for the Bomb*, 324.
108. Overy, *The Air War*, 120.

109. Taylor, *Final Report*, 65.
110. Davis Biddle, "Air Power," 157–58.
111. Thomas, *The Ethics of Destruction*, 148.
112. Davis Biddle, "Air Power," 141.
113. Kennett, *A History of Strategic Bombing*, 185.
114. Thomas, *The Ethics of Destruction*, 148.
115. Thomas, *The Ethics of Destruction*, 149.
116. Thomas, *The Ethics of Destruction*, 151.
117. Davis Biddle, *Rhetoric and Reality*, 297.
118. Davis Biddle, *Rhetoric and Reality*, 297–98.
119. Thomas, *The Ethics of Destruction*, 153.
120. Thomas, *The Ethics of Destruction*, 153–54.
121. Davis Biddle, *Rhetoric and Reality*, 299.
122. Quoted in Thomas, *The Ethics of Destruction*, 157.
123. Renz and Scheipers, "Discrimination in Aerial Bombing," 29–31.
124. Davis Biddle, *Rhetoric and Reality*, 300.
125. Thomas, *The Ethics of Destruction*, 155.
126. Lee, "The Interim Years of Cyberspace," 62; Renz and Scheipers, "Discrimination in Aerial Bombing," 30–32.
127. Finnemore and Sikkink, "International Norm Dynamics," 904.
128. International Committee of the Red Cross, Protocol Additional to the Geneva Conventions of 12 August 1949, and Relating to the Protection of Victims of International Armed Conflicts (Protocol I), 8 June 1977: Protection of the Civilian Population, http://www.icrc.org/applic/ihl/ihl.nsf/Article.xsp?action=openDocument&documentId=4BEBD9920AE0AEAEC12563CD0051DC9E.
129. International Committee of the Red Cross, Protocol Additional to the Geneva Conventions of 12 August 1949.
130. International Committee of the Red Cross, State Parties to Protocol Additional to the Geneva Conventions of 12 August 1949, and Relating to the Protection of Victims of International Armed Conflicts (Protocol I), 8 June 1977, http://www.icrc.org/applic/ihl/ihl.nsf/States.xsp?xp_viewStates=XPages_NORMStatesParties&xp_treatySelected=470.

131. Parks, "Air War and the Law of War," 70–225; Davis Biddle, "Air Power," 159.
132. Karsh, *The Iran–Iraq War*, 1–8, 12–16, 19–82.
133. Wright, *Dreams and Shadows*, 438.
134. Hiro, *The Longest War*, 205.
135. United Nations, United Nations Security Council Resolution 598, January 16, 1987, http://daccess-ods.un.org/access.nsf/Get?Open&DS=S/RES/598%20(1987)&Lang=E&Area=RESOLUTION.
136. Thomas, *The Ethics of Destruction*, 148.
137. Edwin E. Moïse, "Limited War: The Stereotypes," Clemson University, http://www.clemson.edu/caah/history/FacultyPages/EdMoise/limit1.html.
138. Davis Biddle, "Air Power," 158.
139. Thomas, *The Ethics of Destruction*, 158.
140. Thomas, *The Ethics of Destruction*, 158.
141. Thomas, *The Ethics of Destruction*, 162–63.
142. Renz and Scheipers, "Discrimination in Aerial Bombing," 34.
143. "Syrian Jets Bomb Northern City Overrun by Rebels," *Guardian*, March 5, 2013, http://www.theguardian.com/world/2013/mar/05/syrian-jets-bomb-city-rebels.
144. Human Rights Watch, "Death from the Skies: Deliberate and Indiscriminate Air Strikes on Civilians," April 2013, http://www.hrw.org/sites/default/files/reports/syria0413webwcover_1_0.pdf.
145. Cohen, "The Meaning and Future of Air Power," 200.
146. Quoted in Kennett, *A History of Strategic Bombing*, 178.
147. Price, "A Genealogy of the Chemical Weapons Taboo," 90.

4. NUCLEAR WEAPONS

Epigraph: Nina Tannenwald, *The Nuclear Taboo: The United States and the Non-Use of Nuclear Weapons since 1945* (Cambridge: Cambridge University Press, 2008), 184.

1. Carus, "Defining 'Weapons of Mass Destruction.'"
2. U.S. Department of Energy, *The Manhattan Project*.

3. Thomas Schelling, "The Nuclear Taboo," *MIT International Review* (Spring 2007), http://web.mit.edu/mitir/2007/spring/taboo.pdf.
4. Tannenwald, "Stigmatizing the Bomb," 33.
5. Tannenwald, *The Nuclear Taboo*, 79.
6. Norris, *Racing for the Bomb*, 324.
7. Tannenwald, *The Nuclear Taboo*, 80.
8. Katzenstein, Wendt, and Jepperson, "Norms, Identity, and Culture," 134.
9. Tannenwald, *The Nuclear Taboo*, 81.
10. Bernstein, "The Atomic Bombings Reconsidered," 151.
11. Bernstein, "The Atomic Bombings Reconsidered," 151.
12. Leahy, *I Was There*, 440.
13. Rotblat, "Leaving the Bomb Project," 18.
14. Tannenwald, "Stigmatizing the Bomb," 20.
15. Tannenwald, "Stigmatizing the Bomb," 14.
16. Mandelbaum, *The Nuclear Question*, 18.
17. "Declaration on Atomic Bomb by President Truman and Prime Ministers Attlee and King," Nuclear Files.org, November 15, 1945, http://www.nuclearfiles.org/menu/key-issues/nuclear-energy/history/dec-truma-atlee-king_1945-11-15.htm.
18. Rumble, *The Politics of Nuclear Defence*, 285.
19. U.S. Department of State, Office of the Historian, "The Acheson-Lilienthal & Baruch Plans, 1946," http://history.state.gov/milestones/1945-1952/BaruchPlans.
20. U.S. Department of State, Office of the Historian, "The Acheson-Lilienthal & Baruch Plans, 1946."
21. United Nations, "Resolution on the Reports of the First Committee," January 24, 1946, http://daccess-dds-ny.un.org/doc/RESOLUTION/GEN/NR0/032/52/IMG/NR003252.pdf?OpenElement.
22. Andrei Gromyko, "Address by the Soviet Representative to the United Nations Atomic Energy Commission," June 19, 1946, http://fissilematerials.org/library/GromykoPlan1946.pdf.
23. Holloway, *Stalin and the Bomb*.

24. Burns, "Soviet Peace Charade."
25. Guenter, "Stockholm Peace Appeal," 201–2.
26. Guenter, "Stockholm Peace Appeal," 201–2. While hundreds of millions of people allegedly signed the petition, the authenticity of the numbers cited is in question as they claim to include the entire adult population of the Soviet Union.
27. Pugwash, "About Pugwash," http://www.pugwash.org/about.htm.
28. Tannenwald, "Stigmatizing the Bomb," 21.
29. Carey Sublette, "Nuclear Weapons FAQ Section 4.4.1.4: The Teller-Ulam Design," *Nuclear Weapons FAQ*, July 3, 2007, http://nuclearweaponarchive.org/Nwfaq/Nfaq4.html.
30. J. J. Mangano, E. J. Sternglass, J. M. Gould, J. D. Sherman, J. Brown, and W. McDonnell, "Strontium-90 in Newborns and Childhood Disease," *Archives of Environmental Health* 55, no. 4 (2000): 240–44, http://www.ncbi.nlm.nih.gov/pubmed/11005428.
31. Mark Schreiber, "Lucky Dragon's Lethal Catch," *Japan Times*, March 18, 2012, http://www.japantimes.co.jp/life/2012/03/18/general/lucky-dragons-lethal-catch/#.Uv9L1GJdWS0.
32. Tannenwald, "Stigmatizing the Bomb," 21.
33. Tannenwald, "Stigmatizing the Bomb," 22.
34. Rearden, *History of the Office of the Secretary of Defense*, 425–31.
35. Tannenwald, "Stigmatizing the Bomb," 23.
36. Sagan, "Why Do States Build Nuclear Weapons?," 24–26.
37. Tannenwald, "Stigmatizing the Bomb," 24–25.
38. Tannenwald, "Stigmatizing the Bomb," 24.
39. Osgood, *NATO*, chapter 5.
40. United States, "National Intelligence Estimate 100-05-55," 85.
41. Quoted in Gaddis, "The Origins of Self-Deterrence," 145.
42. Quoted in Wittner, *Resisting the Bomb*, 182.
43. Tannenwald, "Stigmatizing the Bomb," 21.
44. Wittner, *Resisting the Bomb*, 31.
45. Rublee, *Nonproliferation Norms*.

46. Finnemore and Sikkink, "International Norm Dynamics," 904.
47. NAM countries are neither aligned for or against any major bloc or nation; in the case of the Cold War, that largely constituted independence from the Soviet Union and the United States.
48. Tannenwald, "Stigmatizing the Bomb," 28–29, 30.
49. The Soviet Union, the United States, and Great Britain had all concurrently adopted atmospheric testing moratoriums in 1958, but these moratoriums were not codified in any legally binding instrument.
50. U.S. Department of State, "Treaty Banning Nuclear Weapon Tests in the Atmosphere, in Outer Space, and Under Water," http://www.state.gov/t/isn/4797.htm.
51. Nuclear Threat Initiative, "Partial Test Ban Treaty: China and the PTBT," http://www.nti.org/db/china/ptbtorg.htm.
52. Arms Control Association, "Threshold Test Ban Treaty (TTBT)," http://www.armscontrol.org/documents/ttbt.
53. Arms Control Association, "Threshold Test Ban Treaty (TTBT)."
54. U.S. Department of State, "Treaty between the United States of America and the Union of Soviet Socialist Republics on the Limitation of Underground Nuclear Weapon Tests (and Protocol Thereto) (TTBT)," http://www.state.gov/t/isn/5204.htm.
55. United Nations, "Ending Nuclear Testing," http://www.un.org/en/events/againstnucleartestsday/history.shtml#a25.
56. Medalia, "Nuclear Testing."
57. "Gallery of U.S. Nuclear Tests," Nuclear Weapons Archive, August 6, 2001, http://nuclearweaponarchive.org/Usa/Tests/index.html.
58. Comprehensive Test Ban Treaty Organization Preparatory Commission, "The Treaty: 1993–1995 Prelude and Formal Negotiations," http://www.ctbto.org/the-treaty/1993-1996-treaty-negotiations/1993-95-prelude-and-formal-negotiations/.
59. Comprehensive Test Ban Treaty Organization Preparatory Commission, "Comprehensive Nuclear-Test-Ban Treaty," http://www.ctbto.org/fileadmin/content/treaty/treaty_text.pdf.

60. James Martin Center for Nonproliferation Studies, "Comprehensive Test Ban Treaty General Description," March 4, 2011, http://cns.miis.edu/inventory/pdfs/ctbt.pdf.
61. Damien LaVera, "Looking Back: The U.S. Senate Vote on Comprehensive Test Ban Treaty," *Arms Control Today*, October 2004, http://www.armscontrol.org/act/2004_10/LookingBack_CTBT.
62. LaVera, "Looking Back."
63. Lawrence Scheinman, "Comprehensive Test Ban Treaty Info Brief," *Monterey Institute of International Studies*, April 2003, http://www.nti.org/e_research/e3_9a.html.
64. Goldschmidt, "The Negotiation of the Non-Proliferation Treaty," 73.
65. Tannenwald, "Stigmatizing the Bomb," 33.
66. Goldschmidt, "The Negotiation of the Non-Proliferation Treaty," 75.
67. Federation of American Scientists, Treaty on the Non-Proliferation of Nuclear Weapons, Nuclear Information Project, http://www.fas.org/nuke/control/npt.
68. United Nations, Office for Disarmament Affairs, Treaty on the Non-Proliferation of Nuclear Weapons (NPT): Overview, http://www.un.org/disarmament/WMD/Nuclear/NPT.shtml.
69. Goldschmidt, "The Negotiation of the Non-Proliferation Treaty," 75.
70. United Nations, Office for Disarmament Affairs, Treaty on the Non-Proliferation of Nuclear Weapons (NPT).
71. Kennedy, "Press Conference," 280.
72. Rublee, *Nonproliferation Norms*, 50.
73. Tannenwald, "Stigmatizing the Bomb," 35.
74. Frey, "Nuclear Weapons as Symbols," 13.
75. William J. Broad, David E. Sanger, and Raymond Bonner, "A Tale of Nuclear Proliferation: How Pakistani Built His Network," *New York Times*, February 12, 2004, http://www.nytimes.com/2004/02/12/world/a-tale-of-nuclear-proliferation-how-pakistani-built-his-network.html?pagewanted=all&src=pm.
76. Schelling, "The Nuclear Taboo."

77. Frederic L. Kirgis, "North Korea's Withdrawal from the Nuclear Nonproliferation Treaty," American Society of International Law Insights, January 2003, http://www.asil.org/insights/volume/8/issue/2/north-koreas-withdrawal-nuclear-nonproliferation-treaty.
78. Shultz, Perry, Kissinger, and Nunn, "A World Free of Nuclear Weapons."
79. Kahn, *Thinking about the Unthinkable*, 245–46.
80. "Statement on Nuclear Weapons by International Generals and Admirals," Nuclear Files.org, December 5, 1996, http://www.nuclearfiles.org/menu/key-issues/ethics/issues/military/statement-by-international-generals.htm.
81. Barack Obama, "Remarks by President Barack Obama in Hradcany Square: April 5, 2009," White House Office of the Press Secretary, http://www.whitehouse.gov/the_press_office/Remarks-By-President-Barack-Obama-In-Prague-As-Delivered/. Earlier that week President Obama and President Dmitry Medvedev of Russia issued a joint statement before the G20 summit in London committing to a world without nuclear weapons.
82. Shultz, Perry, Kissinger, and Nunn, "Toward a Nuclear-Free World."
83. Blechman, *Unblocking the Road to Zero.*
84. Blechman, *Unblocking the Road to Zero*; Center for Defense Information, "The World's Nuclear Arsenals," NuclearFiles.org, January 21, 2009, http://www.nuclearfiles.org/menu/key-issues/nuclear-weapons/issues/capabilities/global-cdi.htm.
85. Blechman, *Unblocking the Road to Zero.*
86. Paul, *The Tradition of Non-Use.*
87. Kahn, "Nuclear Proliferation," 78.
88. Tannenwald, *The Nuclear Taboo*, 250.
89. United Nations, Office for Disarmament Affairs, "Nuclear-Weapon-Free Zones," http://www.un.org/disarmament/WMD/Nuclear/NWFZ.shtml.
90. Tannenwald, *The Nuclear Taboo*, 292.
91. Freedman, "Disarmament and Other Nuclear Norms," 96.
92. Sagan, "Why Do States Build Nuclear Weapons?," 21.

5. EMERGING-TECHNOLOGY WEAPONS

Epigraph: "Will Rogers Has an Idea about Disarmament Plans," Daily Telegram no. 1063 (December 22, 1929), in *Will Rogers' Daily Telegrams*, vol. 2: *The Hoover Years, 1929–1931*, ed. James Smallwood and Steven Gragert (Stillwater: Oklahoma State University Press, 1978).

1. Quoted in Bob Otto, "Operation Allied Force: Bringing Liddell Hart Full Circle," National Defense University National War College, 3, http://www.dtic.mil/cgi-bin/GetTRDoc?AD=ADA442507.
2. Des Moore, "What Is the Greatest Threat: Global Warming or Terrorism?," Institute for Private Enterprise, May 17, 2007, http://www.ipe.net.au/U3A_0508.pdf.
3. Human Rights Watch, "Death from the Skies: Deliberate and Indiscriminate Air Strikes on Civilians," April 2013, http://www.hrw.org/sites/default/files/reports/syria0413webwcover_1_0.pdf.
4. Norris, *Early Gunpowder Artillery.*
5. Hecht, "Top 10."
6. Morgenthau, "Politics among Nations," 197.
7. Katzenstein, Wendt, and Jepperson, "Norms, Identity, and Culture," 127.
8. Thomas, *The Ethics of Destruction*, 100.
9. Guenter, "Stockholm Peace Appeal," 201–2.
10. Katzenstein, Wendt, and Jepperson, "Norms, Identity, and Culture," 127.
11. Douhet, *The Command of the Air*, 61.
12. Katzenstein, Wendt, and Jepperson, "Norms, Identity, and Culture," 127.
13. Spaight, *Air Power*, 220.
14. "Declaration on Atomic Bomb by President Truman and Prime Ministers Attlee and King: November 15, 1945," Nuclear Files, http://www.nuclearfiles.org/menu/key-issues/nuclear-energy/history/dec-truma-atlee-king_1945-11-15.htm.
15. Thomas, *The Ethics of Destruction*, 118.
16. Kennett, *A History of Strategic Bombing*, 67.

17. Thomas, *The Ethics of Destruction*, 99.
18. Thomas, *The Ethics of Destruction*, 100.
19. Rearden, *History of the Office of the Secretary of Defense*, 425–31.
20. Biological and Toxin Weapons Convention, Department of Peace Studies of the University of Bradford, "List of States Parties to the Convention on the Prohibition of the Development, Production and Stockpiling of Bacteriological (Biological) and Toxin Weapons and on Their Destruction as of June 2005," http://www.unog.ch/80256EE600585943/(httpPages)/7BE6CBBEA0477B52C12571860035FD5C?OpenDocument.
21. Sagan, "Why Do States Build Nuclear Weapons?," 74.
22. U.S. Congress, Senate Report 111-377: wmd Prevention and Preparedness Act of 2009, December 17, 2010, https://www.congress.gov/congressional-report/111th-congress/senate-report/377/1.
23. Thomas, *The Ethics of Destruction*, 157; Renz and Scheipers, "Discrimination in Aerial Bombing," 29–31.
24. Carpenter and Moodie, "Industry and Arms Control," 180.
25. Thomas, *The Ethics of Destruction*, 115.
26. Koblentz, *Living Weapons*, 143.
27. "Syria Chemical Attack: What We Know," BBC News, September 24, 2013, http://www.bbc.co.uk/news/world-middle-east-23927399.
28. Tim Cohen, "Obama: It's the World's 'Red Line' on Syria. Senate Panel Backs Military Strike Plan," CNN, September 4, 2013, http://www.cnn.com/2013/09/04/politics/us-syria/index.html.
29. Nick Cumming-Bruce, "Watchdog Says Syria Has Been Cooperative on Weapons," *New York Times*, October 9, 2013, http://www.nytimes.com/2013/10/10/world/middleeast/syria-chemical-weapons.html?_r=0.

6. CYBER WARFARE

Epigraph: James R. Clapper, "Statement for the Record: Worldwide Threat Assessment of the U.S. Intelligence Community," Senate Select Committee on Intelligence, March 12, 2013, http://www.intelligence.senate.gov/130312/clapper.pdf.

1. Clapper, "Statement for the Record," March 12, 2013.

2. Brown and Poellet, "The Customary International Law of Cyberspace," 129–30.
3. Healey, *A Fierce Domain*, 18.
4. James Lewis, "Significant Cyber Events since 2006," Center for Strategic and International Studies, July 11, 2013, http://csis.org/files/publication/140310_Significant_Cyber_Incidents_Since_2006.pdf.
5. Based on author's analysis of Lewis, "Significant Cyber Events since 2006."
6. Rid, "Cyber War Will Not Take Place," 5–32.
7. Other aspects of cyber security, such as cyber crime and cyber espionage, have been examined in the UN the Second Committee and the Economic and Social Council, but these efforts fall outside the scope of this project and focus on a different category of hostile cyber operations, as outlined at the beginning of this chapter.
8. Carr, *Inside Cyber Warfare*, 34.
9. Maurer, "Cyber Norm Emergence," 17.
10. Maurer, "Cyber Norm Emergence," 17.
11. Maurer, "Cyber Norm Emergence," 17.
12. Spencer Ackerman, "NATO Doesn't Yet Know How to Protect Its Networks," *Wired*, February 1, 2012, http://www.wired.com/2012/02/nato-cyber/.
13. Klimburg, *National Cyber Security Framework Manual*, 4.
14. North Atlantic Treaty Organization, "Active Engagement, Modern Defence: Strategic Concept for the Defence and Security of the Members of the North Atlantic Treaty Organisation, adopted by Heads of State and Government at the NATO in Lisbon, November 19–20, 2010," paragraph 12, http://www.nato.int/strategic-concept/pdf/Strat_Concept_web_en.pdf.
15. International Group of Experts, *Tallinn Manual*.
16. Atlantic Council, "Fact Sheet: The Tallinn Manual on the International Law Applicable to Cyber Warfare," March 28, 2013, http://www.atlanticcouncil.org/events/past-events/tallinn-manual-launch-defines-legal-groundwork-for-cyber-warfare.

17. Gili Cohen, "Israeli Expert Seeks Ethics Code for Cyber Warfare," *Haaretz*, January 20, 2014, http://www.haaretz.com/news/diplomacy-defense/1.569450.
18. Schmitt, "International Law in Cyberspace," 14.
19. Schmitt, "International Law in Cyberspace," 14.
20. United States, "International Strategy for Cyberspace: Prosperity, Security, and Openness in a Networked World," May 2011, 9, http://www.whitehouse.gov/sites/default/files/rss_viewer/international_strategy_for_cyberspace.pdf.
21. Cohen, "Israeli Expert Seeks Ethics Code for Cyber Warfare."
22. Lockheed Martin, "Speedy Qubits Lead the Quantum Evolution," October 28, 2013, http://www.lockheedmartin.com/us/news/features/2013/quantum.html?goback=%2egde_4682795_member_5812625891389366275#%21.
23. Lockheed Martin, "Speedy Qubits Lead the Quantum Evolution"; Steven Rich, "NSA Seeks to Build Quantum Computer That Could Crack Most Types of Encryption," *Washington Post*, January 2, 2014, http://www.washingtonpost.com/world/national-security/nsa-seeks-to-build-quantum-computer-that-could-crack-most-types-of-encryption/2014/01/02/8fff297e-7195-11e3-8def-a33011492df2_story.html.
24. Don Clark, "Intel Explains Rare Moore's Law Stumble," *Wall Street Journal*, November 21, 2013, http://blogs.wsj.com/digits/2013/11/21/intel-explains-rare-moores-law-stumble/.
25. Geoffrey Ingersoll, "U.S. Navy: Hackers Jumping the Air Gap Would Disrupt the World Balance of Power," *Business Insider*, November 19, 2013, http://www.businessinsider.com/navy-acoustic-hackers-could-halt-fleets-2013-11.
26. John Markoff, "Brainlike Computers, Learning from Experience," *New York Times*, December 28, 2013, http://www.nytimes.com/2013/12/29/science/brainlike-computers-learning-from-experience.html; Chris Smith, "Brain-like Processors Coming from Qualcomm Next Year," BGR, December 30, 2013, http://bgr.com/2013/12/30/qualcomm-brain-like-processor-2014/.
27. Smith, "Brainlike Processors Coming from Qualcomm Next Year."
28. Singer and Friedman, *Cybersecurity and Cyberwar*, 96–99.

29. Antone Gonsalves, "Stuxnet Creators Defined 21st Century Warfare," *Computer World*, November 21, 2013. http://news.idg.no/cw/art.cfm?id=D106D2FB-9CBD-E42D-6EA37EB5DE2701D3.
30. U.S. National Intelligence Council, "Global Trends 2030: Alternative Worlds," December 2012, iii, 68, http://www.dni.gov/index.php/about/organization/global-trends-2030.
31. Nick Collins, "Cyber Terrorism Is 'Biggest Threat to Aircraft,'" *Telegraph*, December 27, 2013, http://www.telegraph.co.uk/finance/newsbysector/transport/10526620/Cyber-terrorism-is-biggest-threat-to-aircraft.html.
32. Hofmann and Schneckener, "Special Report," 11–12.
33. Ryngaert, "Working Paper."
34. Clapper, "Statement for the Record," March 12, 2013.
35. James R. Clapper, "Statement for the Record: Worldwide Threat Assessment of the U.S. Intelligence Community," Senate Select Committee on Intelligence, January 29, 2014, http://online.wsj.com/public/resources/documents/DNIthreats2014.pdf.
36. "Cyberwar," *Economist*, July 1, 2010, http://www.economist.com/node/16481504.
37. FireEye Inc., *World War C*, 20.
38. Kello, "The Meaning of the Cyber Revolution," 23.
39. Offices of U.S. Congressmen Markey and Waxman, "Electric Grid Vulnerability," 3.
40. Yasmin Tadjdeh, "Fears of Devastating Cyber-Attacks on Electric Grid, Critical Infrastructure Grow," *National Defense Magazine*, October 2013, 24, http://digital.nationaldefensemagazine.org/i/177663/26.
41. Paulo Shakarian, Hansheng Lei, and Roy Lindelauf, "Power Grid Defense against Malicious Cascading Failure," U.S. Military Academy, http://www.westpoint.edu/nsc/siteassets/sitepages/publications/power_grid_def.pdf.
42. Matthew Wald, "As Worries over the Power Grid Rise, a Drill Will Simulate a Knockout Blow," *New York Times*, August 16, 2013, http://www.nytimes.com/2013/08/17/us/as-worries-over-the-power-grid-rise-a-drill-will-simulate-a-knockout-blow.html?_r=2&pagewanted=print&.

43. Healey, *A Fierce Domain*, 86.
44. Bell, "Cut-Price Stuxnet Successors Possible."
45. Jason Healey, "How Emperor Alexander Militarized American Cyberspace," *Foreign Policy*, November 22, 2013, http://www.foreignpolicy.com/articles/2013/11/06/how_emperor_alexander_militarized_american_cyberspace?wp_login_redirect=0#sthash.63nueywc.dEQQ8Uxd.dpbs.
46. Koblentz and Mazanec, "Viral Warfare," 418–34.
47. Lewis and Timlin, *Cybersecurity and Cyberwarfare*; Liff, "Cyberwar," 401–28.
48. Jason Bender, "The Cyberspace Operations Planner," *Small Wars Journal*, November 5, 2013, http://smallwarsjournal.com/jrnl/art/the-cyberspace-operations-planner; Bruce Arnold, "Cold War to Cyber War: Here's How Weapon Exports Are Controlled," *Conversation*, December 8, 2013, http://theconversation.com/cold-war-to-cyber-war-heres-how-weapon-exports-are-controlled-21173.
49. Lewis and Timlin, *Cybersecurity and Cyberwarfare*; Liff, "Cyberwar," 401–28.
50. Lewis and Timlin, *Cybersecurity and Cyberwarfare*, 3.
51. "N. Korea Boosting Cyber Warfare Capabilities," Chosunilbo, November 8, 2013, http://english.chosun.com/site/data/html_dir/2013/11/05/2013110501790.html.
52. William Wan and Ellen Nakashima, "Report Ties Cyberattacks on U.S. Computers to Chinese Military," *Washington Post*, February 19, 2013, http://articles.washingtonpost.com/2013-02-19/world/37166888_1_chinese-cyber-attacks-extensive-cyber-espionage-chinese-military-unit.
53. Anna Mulrine, "China Is a Lead Cyberattacker of U.S. Military Computers, Pentagon Reports," *Christian Science Monitor*, May 18, 2012, http://www.csmonitor.com/USA/Military/2012/0518/China-is-a-lead-cyberattacker-of-US-military-computers-Pentagon-reports.
54. Brian Mazanec, "The Art of (Cyber) War," *Journal of International Security Affairs* 16 (Spring 2009), http://www.securityaffairs.org/issues/2009/16/mazanec.php.

55. U.S. Secretary of State's International Security Advisory Board Task Force, "Draft Report," 1.
56. Gärtner, Hyde-Price, and Reiter, *Europe's New Security Challenges*, 74.
57. Godwin, "Compensating for Deficiencies," 87–118.
58. Michael Pillsbury, *China Debates the Future Security Environment* (Washington, DC: National Defense University Press, 2000), http://www.fas.org/nuke/guide/china/doctrine/pills2/.
59. Liang and Xiangsui, *Unrestricted Warfare*, 145–56.
60. Sun Tzu, *The Art of War* (Radord, VA: Wilder, 2008), http://classics.mit.edu/Tzu/artwar.html.
61. Blasko, "PLA Force Structure," 51–86.
62. FireEye Inc., *World War C*, 7.
63. Liang and Xiangsui, *Unrestricted Warfare*. 10.
64. Richard Lawless, "China: Recent Security Developments," U.S. House Armed Services Committee, June 13, 2007, http://armedservices.house.gov/pdfs/FC061307/Lawless_Testimony061307.pdf.
65. Michael McConnell, "Annual Threat Assessment of the Intelligence Community," U.S. House Armed Services Committee, February 27, 2008, http://www.projectcyw-d.org/resources/items/show/89.
66. U.S.-China Economic and Security Review Commission, "2007 Report to Congress: Conclusion on China's Military Modernization," 110 Cong., 1st sess. USCC.gov, November 2007, 7, http://origin.www.uscc.gov/sites/default/files/annual_reports/2007-Report-to-Congress.pdf.
67. Information Office of the State Council of the People's Republic of China, "China's National Defense in 2004," http://english.peopledaily.com.cn/whitepaper/defense2004/defense2004.html.
68. Information Office of the State Council of the People's Republic of China, "China's National Defense in 2004."
69. Tkacik, "Trojan Dragon."
70. Jaikumar Vijayan, "SANS Sees Upsurge in Zero-Day Web-Based Attacks," *Computer World*, November, 15, 2006, http://www.computerworld.com/action/article.do?command=viewArticleBasic&articleId=9005117.
71. FireEye Inc., *World War C*, 5.

72. U.S.-China Economic and Security Review Commission, "Occupying the Information High Ground: Chinese Capabilities for Computer Network Operations and Cyber Espionage," USCC, March 7, 2012, http://origin.www.uscc.gov/sites/default/files/Research/USCC_Report_Chinese_Capabilities_for_Computer_Network_Operations_and_Cyber_%20Espionage.pdf.
73. Brian Grow et al., "Dangerous Fakes: How Counterfeit, Defective Computer Components from China Are Getting into U.S. Warplanes and Ships," *BusinessWeek*, October 2, 2008, http://www.businessweek.com/magazine/content/08_41/b4103034193886.htm?campaign_id=rss_as.
74. United Nations General Assembly, A/66/359, Letter dated 12 September 2011 from the Permanent Representatives of China, the Russian Federation, Tajikistan, and Uzbekistan to the United Nations addressed to the Secretary-General, September 14, 2011, http://cs.brown.edu/courses/csci1800/sources/2012_UN_Russia_and_China_Code_o_Conduct.pdf.
75. FireEye Inc., *World War C*, 11.
76. Lewis and Timlin, *Cybersecurity and Cyberwarfare*, 19.
77. Heickerö, "Emerging Cyber Threats," 27.
78. Lewis and Timlin, *Cybersecurity and Cyberwarfare*, 19.
79. McConnell, "Annual Threat Assessment of the Intelligence Community"; FireEye Inc., *World War C*, 12.
80. "Being Strong: National Security Guarantees for Russia," Russia Today, February 19, 2012, http://rt.com/politics/official-word/strong-putin-military-russia-711/.
81. FireEye Inc., *World War C*, 12; Peter Bergen and Tim Maurer, "Cyberwar Hits Ukraine," CNN, March 7, 2014, http://www.cnn.com/2014/03/07/opinion/bergen-ukraine-cyber-attacks/; Patrick Tucker, "Weekend Cyberattacks Target NATO, U.S. Military Commands," Defense One, March 17, 2014, http://www.defenseone.com/technology/2014/03/weekend-cyberattacks-target-nato-us-military-commands/80605/?oref=defenseone_today_nl.
82. Lewis, "Significant Cyber Events since 2006," 16.
83. Healey, "How Emperor Alexander Militarized American Cyberspace."
84. FireEye Inc., *World War C*, 17.

85. Lewis and Timlin, *Cybersecurity and Cyberwarfare*, 21.
86. Aliya Sternstein, "Pentagon Plans to Deploy More Than 100 Cyber Teams by Late 2015," NextGov, March 19, 2013, http://www.nextgov.com/defense/2013/03/pentagon-plans-deploy-more-100-cyber-teams-late-2015/61948/.
87. Aliya Sternstein, "Cyber Command Budget More Than Doubles," Defense One, January 14, 2014, http://www.defenseone.com/politics/2014/01/cyber-command-budget-more-doubles/76865/?oref=defenseone_today_nl.
88. Barton Gellman and Ellen Nakashima, "U.S. Spy Agencies Mounted 231 Offensive Cyber-Operations in 2011, Documents Show," *Washington Post*, August 30, 2013, http://articles.washingtonpost.com/2013-08-30/world/41620705_1_computer-worm-former-u-s-officials-obama-administration.
89. United States, "International Strategy for Cyberspace," May 2011, 14; Ellen Nakashima, "In Cyberwarfare, Rules of Engagement Still Hard to Define," *Washington Post*, March 10, 2013, http://www.washingtonpost.com/world/national-security/in-cyberwarfare-rules-of-engagement-still-hard-to-define/2013/03/10/0442507c-88da-11e2-9d71-f0feafdd1394_story.html.
90. William J. Lynn III, "The Pentagon's Cyberstrategy, One Year Later," *Foreign Affairs*, September 28, 2011, http://www.foreignaffairs.com/articles/68305/william-j-lynn-iii/the-pentagons-cyberstrategy-one-year-later.
91. U.S. Department of Defense, "Department of Defense Strategy for Operating in Cyberspace," July 2011, http://www.defense.gov/news/d20110714cyber.pdf.
92. Gorman and Barnes, "Cyber Combat."
93. U.S. Department of Defense, "Defense Science Board Task Force Report: Resilient Military Systems and the Advanced Cyber Threat," January 2013. http://www.acq.osd.mil/dsb/reports/ResilientMilitarySystems.CyberThreat.pdf.
94. Vincent Manzo, "Stuxnet and the Dangers of Cyberwar," *National Interest*, January 29, 2013, http://nationalinterest.org/commentary/stuxnet-the-dangers-cyberwar-8030.

95. Zachary Fryer-Biggs, "DoD Close to Approving Cyber Attack Rules," *Federal Times*, May 28, 2013, http://www.federaltimes.com/article/20130528/it01/305280001/.
96. Kurt Eichenwald, "How Edward Snowden Escalated Cyber War with China," *Newsweek*, November 1, 2013, http://www.newsweek.com/how-edward-snowden-escalated-cyber-war-1461.
97. Gellman and Nakashima, "U.S. Spy Agencies Mounted 231 Offensive Cyber-Operations in 2011, Documents Show."
98. Tero Kuittinen, "NSA Spying Has Triggered a Crazy Chain Reaction of Countermeasure," BGR, December 16, 2013, http://bgr.com/2013/12/16/nsa-spying-finland-fiber/.
99. Kuittinen, "NSA Spying Has Triggered a Crazy Chain Reaction of Countermeasure"; Kathleen Hennessey and Vincent Bevins, "Brazil's President, Angry about Spying, Cancels State Visit to U.S.," *Los Angeles Times*, September 17, 2013, http://articles.latimes.com/2013/sep/17/world/la-fg-snowden-fallout-20130918.
100. David DeWalt, "Going There: The Year Ahead in Cyber Security," Re/Code, February 5, 2014, http://recode.net/2014/02/05/going-there-the-year-ahead-in-cyber-security/.
101. Vivienne Walt, "Greenwald on Snowden Leaks: The Worst Is Yet to Come," *Time*, October 14, 2013, http://world.time.com/2013/10/14/greenwald-on-snowden-leaks-the-worst-is-yet-to-come/.
102. Paul Meyer, "Cyber Security Takes the UN Floor," ICT4Peace Foundation, November 11, 2013, http://ict4peace.org/?p=3000.
103. Bill Gertz, "China Cyber Espionage Grows: Secret Military Cyber Unit Masked Activities after Exposure," *Washington Free Beacon*, November 6, 2013, http://freebeacon.com/china-cyber-espionage-grows/.
104. U.S. National Intelligence Council, "Global Trends 2030," ix.
105. Finnemore, "Cultivating International Cyber Norms," 99.
106. International Group of Experts, *Tallinn Manual*.
107. Brown and Poellet, "The Customary International Law of Cyberspace," 141.

108. Based on author's analysis of Lewis, "Significant Cyber Events since 2006."
109. Elisabeth Bumiller and Thom Shanker, "Panetta Warns of Dire Threat of Cyberattack on U.S.," *New York Times*, October 11, 2012, http://www.nytimes.com/2012/10/12/world/panetta-warns-of-dire-threat-of-cyberattack.html?_r=0.
110. Rid, *Cyber War Will Not Take Place*; Thomas Rid, "Cyberwar and Peace: Hacking Can Reduce Real-World Violence," *Foreign Affairs*, December 2013, http://www.foreignaffairs.com/articles/140160/thomas-rid/cyberwar-and-peace.
111. Gartzke, "The Myth of Cyberwar," 42.
112. Ralph Langner, "Stuxnet's Secret Twin: The Real Program to Sabotage Iran's Nuclear Facilities Was Far More Sophisticated than Anyone Realized," *Foreign Policy*, November 21, 2013, http://foreignpolicy.com/2013/11/19/stuxnets-secret-twin/.
113. John Style and Angie Petty, "Federal Information Security Market, 2010–2015," Deltek, November 2010, http://iq.govwin.com/corp/library/detail.cfm?ItemID=13648.
114. Bruce Schneier, "Threat of 'Cyberwar' Has Been Hugely Hyped," CNN, July 7, 2010, http://www.cnn.com/2010/OPINION/07/07/schneier.cyberwar.hyped/.
115. Jerry Brito and Tate Watkins, "Loving the Cyber Bomb? The Dangers of Threat Inflation in Cybersecurity Policy," Mercatus Center, George Mason University, April 26, 2011, http://mercatus.org/publication/loving-cyber-bomb-dangers-threat-inflation-cybersecurity-policy; "Is Cyberwar Hype Fuelling a Cybersecurity-Industrial Complex?," Russia Today, February 16, 2012, http://rt.com/usa/security-us-cyber-threat-529/.
116. U.S. Department of Defense, "Defense Science Board Task Force Report," 1.
117. Nicole Perlroth, "In Cyberattack on Saudi Firm, U.S. Sees Iran Firing Back," *New York Times*, October 23, 2012, http://www.nytimes

.com/2012/10/24/business/global/cyberattack-on-saudi-oil-firm-disquiets-us.html.

118. As evidenced by the examination of customary state practice of cyber warfare in the appendix, which identified these as primary cyber actors.
119. Hathaway, "Cyber Readiness Index 1.0."
120. Jack Goldsmith, "The Significance of Panetta's Cyber Speech and the Persistent Difficulty of Deterring Cyberattacks," Lawfare, October 15, 2012, http://www.lawfareblog.com/2012/10/the-significance-of-panettas-cyber-speech-and-the-persistent-difficulty-of-deterring-cyberattacks/.
121. Knake, "Testimony."
122. Geoffrey Ingersoll, "The Next Weapon of Mass Destruction Will Probably Be a Thumbdrive," *Business Insider*, October 29, 2012, http://www.businessinsider.com/the-next-weapon-of-mass-destruction-will-probably-be-a-thumbdrive-2012-10.
123. Jeffrey Carr, "The Misunderstood Acronym: Why Cyber Weapons Aren't WMD," *Bulletin of the Atomic Scientists*, October 17, 2013, http://www.thebulletin.org/2013/september/misunderstood-acronym-why-cyber-weapons-arent-wmd.
124. Tom Brewster, "Governments to Show 'Public Demonstrations of Cyber War Power,'" Tech Week Europe, November 1, 2012, http://www.techweekeurope.co.uk/interview/cyber-war-mikko-hypponen-97983.
125. "Attachment II: Best Practices," 2013 Seoul Cyber Conference, October 18, 2013, http://www.mofa.go.kr/webmodule/common/download.jsp?boardid=14472&tablename=TYPE_ENGLEGATIO&seqno=f84050f95fe8024003fdffc1&fileseq=023feafe6038009fb3051ffb.
126. Eva Galperin, Morgan Marquis-Boire, and John Scott-Railton, "Quantum of Surveillance: Familiar Actors and Possible False Flags in Syrian Malware Campaigns," Citizen Lab and Electronic Frontier Foundation, December 23, 2013, https://www.eff.org/document/quantum-surveillance-familiar-actors-and-possible-false-flags-syrian-malware-campaigns.
127. Eichenwald, "How Edward Snowden Escalated Cyber War with China."

128. Barton Gellman and Laura Poitras, "U.S., British Intelligence Mining Data from Nine U.S. Internet Companies in Broad Secret Program," *Washington Post*, June 6, 2013, http://www.washingtonpost.com/investigations/us-intelligence-mining-data-from-nine-us-internet-companies-in-broad-secret-program/2013/06/06/3a0c0da8-cebf-11e2-8845-d970ccb04497_story.html.
129. Margaret Talev and Chris Strohm, "NSA Fallout Tests Obama Relationship with Tech Companies," *Bloomberg News*, December 18, 2013, http://www.bloomberg.com/news/2013-12-18/nsa-fallout-tests-obama-relationship-with-tech-companies.html.
130. Greg Miller, "FBI Director Warns of Cyberattacks: Other Security Chiefs Say Terrorism Threat Has Altered," *Washington Post*, November 14, 2013, http://www.washingtonpost.com/world/national-security/fbi-director-warns-of-cyberattacks-other-security-chiefs-say-terrorism-threat-has-altered/2013/11/14/24f1b27a-4d53-11e3-9890-a1e0997fb0c0_story.html.
131. Patrick Thibodeau, "Cyberattacks Could Paralyze U.S., Former Defense Chief Warns," *ComputerWorld*, March 11, 2014, http://www.computerworld.com/s/article/9246886/Cyberattacks_could_paralyze_U.S._former_defense_chief_warns.

CONCLUSIONS

1. Zachary Fryer-Biggs, "Poll: Cyberwarfare Is Top Threat Facing U.S.," *Defense News*, January 5, 2014, http://www.defensenews.com/article/20140105/DEFREG02/301050011/Poll-Cyberwarfare-top-Threat-Facing-US?odyssey=nav%7Chead; Bruce Stokes, "Extremists, Cyber-Attacks Top Americans' Security Threat List," Pew Research Center, January 2, 2014, http://www.pewresearch.org/fact-tank/2014/01/02/americans-see-extremists-cyber-attacks-as-major-threats-to-the-u-s/.
2. James R. Clapper, "Statement for the Record: Worldwide Threat Assessment of the U.S. Intelligence Community," Senate Select Committee on Intelligence, March 12, 2013, http://online.wsj.com/public/resources/documents/DNIthreats2014.pdf.

3. Victor Mallet, "Mutually Assured Destruction in Cyberspace," *Financial Times*, August 20, 2008, http://www.ft.com/cms/s/0/ca5cb050-6eb7-11dd-a80a-0000779fd18c.html#axzz3YBAbUJGP; Clarke and Knake, *Cyber War*, 240–51, 268.
4. Geoffrey Ingersoll, "America May Have Opened the Pandora's Box of Cyberwarfare," *Business Insider*, November 21, 2013, http://www.seattlepi.com/technology/businessinsider/article/America-May-Have-Opened-The-Pandora-s-Box-Of-4999927.php; Matthew Wald, "As Worries over the Power Grid Rise, a Drill Will Simulate a Knockout Blow," *New York Times*, August 16, 2013, http://www.nytimes.com/2013/08/17/us/as-worries-over-the-power-grid-rise-a-drill-will-simulate-a-knockout-blow.html?_r=2&pagewanted=print&.
5. Vincent Manzo, "Stuxnet and the Dangers of Cyberwar," *National Interest*, January 29, 2013, http://nationalinterest.org/commentary/stuxnet-the-dangers-cyberwar-8030.
6. Nicholas Hoover, "Defense Bill Approves Offensive Cyber Warfare," *Information Week*, January 5, 2012, http://www.informationweek.com/security/risk-management/-defense-bill-approves-offensive-cyber-warfare/d/d-id/1102072?.
7. "Cyberwar, High-Tech Weapons Take Center Stage in Defense Budget," Fox News, December 27, 2013, http://www.foxnews.com/tech/2013/12/27/cyberwar-high-tech-weapons-take-center-stage-in-defense-budget/.
8. Ralph Langner, "Stuxnet's Secret Twin: The Real Program to Sabotage Iran's Nuclear Facilities Was Far More Sophisticated than Anyone Realized," *Foreign Policy*, November 21, 2013, http://foreignpolicy.com/2013/11/19/stuxnets-secret-twin/.
9. Noah Shachtman, "John Arquilla: Go on the Cyberoffensive," *Wired*, September 21, 2009, http://www.wired.com/techbiz/people/magazine/17-10/ff_smartlist_arquilla.
10. Based on a presentation given by Jason Healey on December 16, 2013, at the Johns Hopkins University Applied Physics Laboratory, Rethinking U.S. Enduring Strengths, Challenges, and Opportunities seminar series.

11. "Japan Seeks Cyberwarfare Capability: Tokyo Looks to U.S. for Ways to Keep Computer Systems Safe," *Japan Times*, December 22, 2013, http://www.japantimes.co.jp/news/2013/12/22/national/japan-seeks-cyberwarfare-capability/#.UsFUu9JDuSr.
12. Charles Miranda, "UK Develops Cyber Strike Capability," *News.com*, December 25, 2013, http://www.news.com.au/world/uk-develops-cyber-strike-capability/story-fndir2ev-1226789746476.
13. Arquilla, "Twenty Years of Cyberwar," 84–85.
14. Lewis, "Confidence-Building," 56.
15. United Nations General Assembly, A/68/98, "Report of the Group of Governmental Experts," 9; ICT4Peace, "ICT4Peace Special Session: Norms and CBMs. Moving towards a More Inclusive Agenda, Summary Report," October 17, 2013, http://mercury.ethz.ch/serviceengine/Files/ISN/172917/ipublicationdocument_singledocument/082e6a53-16ee-4218-b643-4b8fc965bdb0/en/ICT4+Peace+Report+plus+Statement+Seoul+-+Norms+and+CBMs+-+final.pdf.
16. Gady and Austin, *Russia, the United States, and Cyber Diplomacy*, acknowledgments.
17. James Lewis, "Significant Cyber Events since 2006," Center for Strategic and International Studies, July 11, 2013, http://csis.org/files/publication/140310_Significant_Cyber_Incidents_Since_2006.pdf.
18. Miranda, "UK Develops Cyber Strike Capability."
19. "Iran's Supreme Leader Tells Students to Prepare for Cyber War," Russia Today, February 13, 2014, http://rt.com/news/iran-israel-cyber-war-899/.
20. Hecht, "Top 10."
21. Elster, "Social Norms and Economic Theory," 99–117.

APPENDIX

1. Woolley, "Defining Cyberspace," 2–3.
2. U.S. Department of Defense, *Joint Publication 1-02*, 93.
3. U.S. Department of Defense, "The Definition of Cyberspace."
4. U.S. Department of Homeland Security, "U.S. National Strategy to Secure Cyberspace," 2003, http://www.dhs.gov/national-strategy-secure-cyberspace.

5. International Telecommunications Union, "2010 U.S. Internet Usage and Broadband Report," 2011, http://www.itu.int/en/ITU-D/Statistics/Documents/facts/ICTFactsFigures2010.pdf.
6. United States, "International Strategy for Cyberspace: Prosperity, Security, and Openness in a Networked World," May 2011, http://www.whitehouse.gov/sites/default/files/rss_viewer/international_strategy_for_cyberspace.pdf.
7. International Telecommunications Union, "The World in 2011—ICT Facts and Figures," December 2011, http://www.itu.int/en/ITU-D/Statistics/Documents/facts/ICTFactsFigures2011.pdf.
8. Ackerman, "NATO Doesn't Yet Know How to Protect Its Networks."
9. U.S. Department of Homeland Security, "U.S. National Strategy to Secure Cyberspace."
10. Tom Gjelten, "Cyber Insecurity: U.S. Struggles to Confront Threat," NPR, April 6, 2012, http://www.npr.org/templates/story/story.php?storyId=125578576.
11. Arquilla, "Twenty Years of Cyberwar," 85.
12. David E. Sanger and Steven Erlanger, "Suspicion Falls on Russia as 'Snake' Cyberattacks Target Ukraine's Government," *New York Times*, March 8, 2014, http://www.nytimes.com/2014/03/09/world/europe/suspicion-falls-on-russia-as-snake-cyberattacks-target-ukraines-government.html?_r=0.
13. Sanger and Erlanger, "Suspicion Falls on Russia."
14. Sanger and Erlanger, "Suspicion Falls on Russia."
15. Brigid Grauman, "Cyber-Security: The Vexed Question of Global Rules," Security Defence Agenda and McAfee, February 2012, 6, http://www.mcafee.com/au/resources/reports/rp-sda-cyber-security.pdf.
16. Jeremy Bender, "Israel: Cyber Is a Bigger Revolution in Warfare Than Gunpowder," *Business Insider*, February 4, 2014, http://www.businessinsider.com/the-internet-is-the-next-battlefield-2014-2.
17. Lewis and Timlin, "Cybersecurity and Cyberwarfare"; Liff, "Cyberwar," 401–28.
18. United States, "2007 Report to Congress: Conclusion on China's Military Modernization," 110 Cong., 1st sess. USCC.gov, November

2007, http://rigin.www.uscc.gov/sites/default/files/annual_reports/2007-Report-to-Congress.pdf, 94.
19. U.S. Department of Defense, "DOD Information Operations Roadmap," October 30, 2003, http://www.gwu.edu/~nsarchiv/NSAEBB/NSAEBB177/info_ops_roadmap.pdf.
20. U.S. Department of Defense, "2014 Quadrennial Defense Review," http://www.defense.gov/pubs/2014_Quadrennial_Defense_Review.pdf.
21. U.S. Department of Defense, *Joint Publication 3-13.1: Electronic Warfare*, January 2007, www.dtic.mil/doctrine/jel/new_pubs/jp3_13.1.pdf.
22. Horowitz, *The Diffusion of Military Power.*
23. Koblentz and Mazanec, "Viral Warfare," 418–34.
24. Rid, "Cyber War Will Not Take Place," 5–32.
25. François Paget, "How Many Bot-Infected Machines on the Internet?," McAfee Labs, January 29, 2007, http://blogs.mcafee.com/mcafee-labs/how-many-bot-infected-machines-are-on-the-internet.
26. SRI International, *An Analysis of Conficker's Logic and Rendezvous Points*, March 19, 2009, http://mtc.sri.com/Conficker/.
27. Lyons, "Threat Assessment of Cyber Warfare," 17.
28. Entous and Hodge, "Oil Firms Hit."
29. U.S. National Research Council, *Technology, Policy, Law, and Ethics*, 82.
30. Webber, "Computer Use Expected to Top 2 Billion."
31. Burton, "The Conficker Worm."
32. James Lewis, "The 'Korean' Cyber Attacks and Their Implications for Cyber Conflict," Center for Strategic and International Studies (October 2009), http://csis.org/publication/korean-cyber-attacks-and-their-implications-cyber-conflict.
33. Lewis and Timlin, *Cybersecurity and Cyberwarfare*, 3.
34. Koblentz and Mazanec, "Viral Warfare," 418–34.
35. Lewis and Timlin, *Cybersecurity and Cyberwarfare*, 3.
36. Lewis and Timlin, *Cybersecurity and Cyberwarfare*, 3; Peter Judge, "Quantum Computer to Debut Next Week," *Tech World*, February 8, 2007, http://www.techworld.com/opsys/news/index.cfm?newsid=7972; Lockheed Martin, "Speedy Qubits Lead the Quantum Evolution," October 28,

2013, http://www.lockheedmartin.com/us/news/features/2013/quantum.html?goback=%2egde_4682795_member_5812625891389366275#%21. A qubit is a quantum bit, the quantum computing equivalent to the binary digit or bit in classical computing.

37. Kopp, "Moore's Law."
38. Bell, "Cut-Price Stuxnet Successors Possible."
39. U.S. General Accounting Office, "B-2 Bomber."
40. Liff, "Cyberwar," 401–28.
41. Bell, "Cut-Price Stuxnet Successors Possible."
42. Denning, *Information Warfare and Security*, 17.
43. James C. Mulvenon, "China's Proliferation Practices and the Development of Its Cyber and Space Warfare Capabilities," U.S.-China Economic and Security Review Commission Hearing, May 20, 2008, 70, http://www.uscc.gov/sites/default/files/transcripts/5.20.08HearingTranscript.pdf.
44. U.S. National Research Council, *Technology, Policy, Law, and Ethics*, 87.
45. Denning, *Information Warfare and Security*, 17.
46. Cyber Conflict Studies Association, "Implication for an Estonia-Like Cyber Conflict for the Government and the Private Sector."
47. Liang and Xiangsui, *Unrestricted Warfare*.
48. Pape, *Bombing to Win*, 72.
49. Cliff et al., *Entering the Dragon's Lair*, 55.
50. Houqing and Zhang, *Science of Campaigns*.
51. Brown and Poellet, "The Customary International Law of Cyberspace," 126.
52. Brown and Poellet, "The Customary International Law of Cyberspace," 130.
53. Paget, "How Many Bot-Infected Machines on the Internet?"
54. Brown and Poellet, "The Customary International Law of Cyberspace," 130.
55. Gus W. Weiss, "The Farewell Dossier: Duping the Soviets," Central Intelligence Agency, June 27, 2008, https://www.cia.gov/library/center-for-the-study-of-intelligence/csi-publications/csi-studies/studies/96unclass/farewell.htm.

56. For disputes over the veracity of the reports regarding the attack, see Jeffrey Carr, "The Myth of the CIA and the Trans-Siberian Pipeline Explosion," Digital Dao, June 7, 2012, http://jeffreycarr.blogspot.com/2012/06/myth-of-cia-and-trans-siberian-pipeline.html. For information on the alleged effects of the attack, see Brown and Poellet, "The Customary International Law of Cyberspace," 130.
57. Steve Melito, "Cyber War and the Siberian Pipeline Explosion," CBRN Resource Network, November 2, 2013, http://www.dsalert.org/int-experts-opinion/cyber-warfare/508-cyber-war-and-the-siberian-pipeline-explosion.
58. Larry Greenemeier, "Estonian 'Cyber Riot' Was Planned, but Mastermind Still a Mystery," *Information Week*, August 3, 2007, http://www.informationweek.com/estonian-cyber-riot-was-planned-but-mast/201202784.
59. North Atlantic Treaty Organization Cooperative Cyber Defence Center of Excellence website, https://www.ccdcoe.org/.
60. Lewis, "The 'Korean' Cyber Attacks."
61. John Markoff, "Before the Gunfire, Cyberattacks," *New York Times*, August 13, 2008, http://www.nytimes.com/2008/08/13/technology/13cyber.html?_r=0; David Hollis, "Cyberwar Case Study: Georgia 2008," *Small Wars Journal*, January 6, 2011, http://smallwarsjournal.com/jrnl/art/cyberwar-case-study-georgia-2008.
62. Gregg Keizer, "Georgian Cyberattacks Suggest Russian Involvement," *ComputerWorld*, October 17, 2008, http://www.computerworld.com/s/article/9117439/Georgian_cyberattacks_suggest_Russian_involvement_say_researchers.
63. Kim Zetter, "How Digital Detectives Deciphered Stuxnet, the Most Menacing Malware in History," *Wired*, July 11, 2011, http://www.wired.com/threatlevel/2011/07/how-digital-detectives-deciphered-stuxnet/.
64. Zetter, "How Digital Detectives Deciphered Stuxnet," 2.
65. David Albright, Paul Brannan, and Christina Walrond, "Stuxnet Malware and Natanz," Institute for Science and International Security, February 15, 2011, http://isis-online.org/isis-reports/detail/stuxnet-malware-and-natanz-update-of-isis-december-22-2010-reportsupa-href1/.

66. Paulo Shakarian, "Stuxnet: Cyberwar Revolution in Military Affairs," *Small Wars Journal*, April 14, 2011, 1, http://smallwarsjournal.com/jrnl/art/stuxnet-cyberwar-revolution-in-military-affairs.
67. Shakarian, "Stuxnet," 1.
68. Shakarian, "Stuxnet," 6. Zero-day vulnerabilities refer to previously unrecognized vulnerabilities in software code. Soon after they are exploited, they are often patched by the software developer, eliminating the vulnerability.
69. Shakarian, "Stuxnet," 1.
70. David Sanger, "Obama Ordered Sped Up Wave of Cyberattacks against Iran," *New York Times*, June 1, 2012, http://www.nytimes.com/2012/06/01/world/middleeast/obama-ordered-wave-of-cyberattacks-against-iran.html?pagewanted=all.
71. Ralph Langner, "Stuxnet's Secret Twin: The Real Program to Sabotage Iran's Nuclear Facilities Was Far More Sophisticated than Anyone Realized," *Foreign Policy*, November 21, 2013, http://foreignpolicy.com/2013/11/19/stuxnets-secret-twin/.
72. Brown and Poellet, "The Customary International Law of Cyberspace," 132.
73. "Stuxnet 'Badly Infected' Russian Nuclear Plant, Claims Kaspersky," Graham Cluley, November 10, 2013, http://grahamcluley.com/2013/11/stuxnet-badly-infected-russian-nuclear-plant-claims-kaspersky/?utm_source=rss&utm_medium=rss&utm_campaign=stuxnet-badly-infected-russian-nuclear-plant-claims-kaspersky.
74. Nicole Perlroth, "In Cyberattack on Saudi Firm, U.S. Sees Iran Firing Back," *New York Times,* October 23, 2012, http://www.nytimes.com/2012/10/24/business/global/cyberattack-on-saudi-oil-firm-disquiets-us.html.
75. Bronk and Tikk-Ringas, "The Cyber Attack on Saudi Aramco," 81–96.
76. Wael Mahdi, "Saudi Arabia Says Aramco Cyberattack Came from Foreign States," *Bloomberg News*, December 9, 2012, http://www.bloomberg.com/news/2012-12-09/saudi-arabia-says-aramco-cyberattack-came-from-foreign-states.html.

77. Mahdi, "Saudi Arabia Says Aramco Cyberattack Came from Foreign States."
78. James Lewis, "Significant Cyber Events since 2006," Center for Strategic and International Studies, July 11, 2013, 12, http://csis.org/files/publication/140310_Significant_Cyber_Incidents_Since_2006.pdf.
79. Bronk and Tikk-Ringas, "The Cyber Attack on Saudi Aramco," 81–96.
80. Perlroth, "In Cyberattack on Saudi Firm, U.S. Sees Iran Firing Back."
81. Lewis, "Significant Cyber Events since 2006," 12–13.
82. Mathew J. Schwartz, "Bank Attackers Restart Operation Ababil DDoS Disruptions," *InformationWeek Security*, March 6, 2013, http://www.informationweek.com/security/attacks/bank-attackers-restart-operation-ababil/240150175.
83. Lewis, "Significant Cyber Events since 2006," 12–13.
84. Ellen Nakashima, "Iran Blamed for Cyberattacks on U.S. Banks and Companies," *Washington Post*, September 21, 2012, http://articles.washingtonpost.com/2012-09-21/world/35497878_1_web-sites-quds-force-cyberattacks.
85. John Markoff, and Thom Shanker, "Halted '03 Plan Illustrates U.S. Fear of Cyber Risk," *New York Times*, August 1, 2009, http://www.nytimes.com/2009/08/02/us/politics/02cyber.html.
86. Ellen Nakashima, "U.S. Cyberweapons Had Been Considered to Disrupt Gaddafi's Air Defenses," *Washington Post*, October 17, 2011, http://www.washingtonpost.com/world/national-security/us-cyber-weapons-had-been-considered-to-disrupt-gaddafis-air-defenses/2011/10/17/gIQAETpssL_story.html.
87. Joseph Menn, "Syria, Aided by Iran, Could Strike Back at U.S. in Cyberspace," Reuters, August 29, 2013, http://www.reuters.com/article/2013/08/29/us-syria-crisis-cyberspace-analysis-idUSBRE97S04Z20130829.
88. Healey, *A Fierce Domain*, 23.
89. Brown and Poellet, "The Customary International Law of Cyberspace," 132.
90. Council of Europe, Convention on Cybercrime, November 23, 2011, http://conventions.coe.int/Treaty/en/Treaties/Html/185.htm.
91. Carr, *Inside Cyber Warfare*, 63.

92. United Nations, "Basic Information: About WSIS," http://www.itu.int/wsis/basic/about.html.
93. United Nations, Charter of the United Nations and Statute of the International Court of Justice, Article 2, paragraph 4, http://treaties.un.org/doc/Publication/CTC/uncharter.pdf.
94. The Second Committee, the Economic and Social Council, and other UN organizations have looked at other aspects of cyber security, such as cyber crime and cyber espionage, but these efforts fall outside the scope of this project.
95. Carr, *Inside Cyber Warfare*, 34.
96. Clarke and Knake, *Cyber War*, 220.
97. Maurer, "Cyber Norm Emergence at the United Nations," 20.
98. Maurer, "Cyber Norm Emergence at the United Nations," 22.
99. Maurer, "Cyber Norm Emergence at the United Nations," 22.
100. Maurer, "Cyber Norm Emergence at the United Nations," 6.
101. Maurer, "Cyber Norm Emergence at the United Nations," 26; UN General Assembly Resolutions and Decisions Database.
102. Maurer, "Cyber Norm Emergence at the United Nations," 21.
103. United Nations General Assembly, A/66/359, Letter dated 12 September 2011 from the Permanent Representatives of China, the Russian Federation, Tajikistan, and Uzbekistan to the United Nations addressed to the Secretary-General, September 14, 2011, https://ccdcoe.org/sites/default/files/documents/UN-110912-CodeOfConduct_0.pdf.
104. Maurer, "Cyber Norm Emergence at the United Nations," 25.
105. Paul Meyer, "Cyber Security Takes the UN Floor," ICT4Peace Foundation, November 11, 2013, http://ict4peace.org/?p=3000.
106. United Nations Office for Disarmament Affairs, "Fact Sheet."
107. United Nations General Assembly, A/68/98, "Report of the Group of Governmental Experts," 2–3, 9.
108. Grauman, "Cyber-Security," 5.
109. ICT4Peace, "ICT4Peace Special Session: Norms and CBMs. Moving towards a More Inclusive Agenda, Summary Report," October 17, 2013, http://mercury.ethz.ch/serviceengine/Files/ISN/172917

/ipublicationdocument_singledocument/082e6a53-16ee-4218-b643-4b8fc965bdb0/en/ICT4+Peace+Report+plus+Statement+Seoul+-+Norms+and+CBMs+-+final.pdf.

110. Maurer, "Cyber Norm Emergence at the United Nations," 9–10.
111. Maurer, "Cyber Norm Emergence at the United Nations," 29.
112. Wegener, "Cyber Peace," 81.
113. Wegener, "Cyber Peace," 81.
114. Maurer, "Cyber Norm Emergence at the United Nations," 28.
115. United Nations Counter Terrorism Implementation Task Force. "Report on Countering the Use of the Internet for Terrorist Purposes," February 2009, 26, http://www.un.org/en/terrorism/ctitf/pdfs/ctitf_internet_wg_2009_report.pdf.
116. Ackerman, "NATO Doesn't Yet Know How to Protect Its Networks."
117. Klimburg, *National Cyber Security Framework Manual*, 4.
118. North Atlantic Treaty Organization, "Active Engagement, Modern Defence: Strategic Concept for the Defence and Security of the Members of the North Atlantic Treaty Organisation, Adopted by Heads of State and Government at the NATO in Lisbon, November 19–20, 2010," paragraph 12, http://www.nato.int/strategic-concept/pdf/Strat_Concept_web_en.pdf.
119. Anna Mulrine, "'American Blackout': Is National Geographic's Take on Cyberattack Accurate?," *Christian Science Monitor*, October 27, 2013, http://www.csmonitor.com/World/Security-Watch/2013/1027/American-Blackout-Is-National-Geographic-s-take-on-cyberattack-accurate-video.
120. Atlantic Council, "Fact Sheet: The Tallinn Manual on the International Law Applicable to Cyber Warfare," March 28, 2013, http://www.atlanticcouncil.org/events/past-events/tallinn-manual-launch-defines-legal-groundwork-for-cyber-warfare.
121. Gili Cohen, "Israeli Expert Seeks Ethics Code for Cyber Warfare," *Haaretz*, January 20, 2014, http://www.haaretz.com/news/diplomacy-defense/1.569450.
122. Schmitt, "International Law in Cyberspace," 14.

123. Schmitt, "International Law in Cyberspace," 14.
124. United States, "International Strategy for Cyberspace," 9.
125. Keith Alexander, "Responses to Advance Questions for Lieutenant General Keith Alexander, USA, Nominee for Commander, United States Cyber Command," April 15, 2010, http://epic.org/Alexander_04-15-10.pdf.
126. Schmitt, "International Law in Cyberspace," 17.
127. See International Group of Experts, *Tallinn Manual*; Schmitt, "International Law in Cyberspace," 17–32.
128. Schmitt, "International Law in Cyberspace," 19; International Group of Experts, *Tallinn Manual* 195.
129. Schmitt, "International Law in Cyberspace," 27.
130. For example, see "NATO Cyberwar Directive Declares Hackers Military Targets," Russia Today, March 19, 2013, http://rt.com/usa/nato-publishes-cyberwar-guidelines-502/; Aaron Souppouris, "Killing Hackers Is Justified in Cyber Warfare, Says NATO-Commissioned Report," *Verge*, March 21, 2013, http://www.theverge.com/2013/3/21/4130740/tallin-manual-on-the-international-law-applicable-to-cyber-warfare.
131. Roger Hurwitz, "An Augmented Summary of the Harvard, MIT and U. of Toronto Cyber Norms Workshop," May 2012, 12, https://www.yumpu.com/en/document/view/18791277/cyber-norms-workshop-ecir-mit.
132. U.K. Foreign and Commonwealth Office, "London Conference on Cyberspace: Chair's Statement," November 2, 2011, https://www.gov.uk/government/news/london-conference-on-cyberspace-chairs-statement.
133. "2012 Budapest Conference," 2013 Seoul Cyber Conference, http://www.mofa.go.kr/english/visa/images/res/Cy_Eng.pdf; "Overview," 2013 Seoul Cyber Conference, http://www.mofa.go.kr/english/visa/images/res/Cy_Eng.pdf.
134. "Citizen Lab Fellows Tim Maurer and Camino Kavanagh on the 2013 Seoul Conference on Cyberspace," Munk School of Global Affairs, University of Toronto, October 28, 2013, https://citizenlab.org/2013/10/citizen-lab-fellows-tim-maurer-camino-kavanagh-2013-seoul-conference-cyberspace/.

135. "2013, Seoul Framework for and Commitment to Open and Secure Cyberspace," 2013 Seoul Cyber Conference, October 18, 2013, http://www.mofat.go.kr/english/visa/images/res/SeoulFramework.pdf.
136. "Attachment II: Best Practices," 2013 Seoul Cyber Conference, October 18, 2013, http://www.mofa.go.kr/webmodule/common/download.jsp?boardid=14468&tablename=TYPE_DATABOARD&seqno=02ff92f82feb07ff96fa3074&fileseq=ffaf8ef8f009079013fa1fe9.
137. "UK Contribution to the Seoul Conference on Cyberspace, Next Steps: Key Activities to Take Forward the London Agenda," 2013 Seoul Cyber Conference, October 18, 2013, 3, https://www.google.com/url?sa=t&rct=j&q=&esrc=s&frm=1&source=web&cd=1&ved=0CB4QFjAA&url=https%3A%2F%2Fwww.gov.uk%2Fgovernment%2Fuploads%2Fsystem%2Fuploads%2Fattachment_data%2Ffile%2F251569%2FNext_Steps.docx&ei=t5Q5VaqfBPj7sATH6YHADg&usg=AFQjCNEV3lx7mQA6-hgldx_Atg5vk6WEow.
138. United Nations Office for Disarmament Affairs, Biological Weapons Convention, http://www.un.org/disarmament/WMD/Bio/.
139. Kim Eun-jung, "Seoul Forum Calls for International Rules for Cyber Warfare," Yonhap News Agency, November 12, 2013, http://english.yonhapnews.co.kr/news/2013/11/12/0200000000AEN20131112007100315F.html.
140. Eun-jung, "Seoul Forum Calls for International Rules for Cyber Warfare."
141. "Attachment II: Best Practices."
142. "S. Korea, U.S. Hold Working-Level Talks on Cybersecurity," Global Post, February 7, 2014, http://www.globalpost.com/dispatch/news/kyodo-news-international/140207/s-korea-us-hold-working-level-talks-cybersecurity.
143. Roger Hurwitz, "A Preliminary Report on the Cyber Norms Workshop," October 2011, https://citizenlab.org/cybernorms2012/introduction.pdf.
144. Hurwitz, "An Augmented Summary of the Harvard, MIT and U. of Toronto Cyber Norms Workshop."

145. Roger Hurwitz, "Introduction to a Preliminary Report on the Harvard, MIT and U. of Toronto Cyber Norms Workshop 2.0," October 3, 2012, http://citizenlab.org/cybernorms2012/introduction.pdf.

146. Kenneth Stewart, "NPS Professors Awarded Prestigious Grant to Explore Ethics of Cyberwarfare," DVIDS, November 14, 2013, http://www.dvidshub.net/news/116755/nps-professors-awarded-prestigious-grant-explore-ethics-cyberwarfare#.UoYMcXCsg6Y; Kathy Jennings, "WMU Prof Helps Define How Cyberwar Can Be Fought Ethically," *Second Wave Media*, October 24, 2013, http://swmichigan.secondwavemedia.com/devnews/ethicalcyberwarfare1024.aspx.

147. Gady and Austin, *Russia, the United States, and Cyber Diplomacy*, acknowledgments, 3.

148. EastWest Institute, *Building Trust in Cyberspace: 3rd Worldwide Cybersecurity Summit in New Delhi*, 2012, http://www.ewi.info/idea/building-trust-cyberspace; Rauscher and Yaschenko, *Russia-U.S. Bilateral on Cybersecurity*.

149. U.S. National Research Council, *Technology, Policy, Law, and Ethics*, 251.

BIBLIOGRAPHY

Note: All sources accessed online are cited in full in the notes and do not appear in the bibliography.

Acharya, Amitav. "How Ideas Spread: Whose Norms Matter? Norm Localization and Institutional Change in Asian Regionalism." *International Organization* 58, no. 2 (2004): 239–75.

Ackerman, Spencer. "NATO Doesn't Yet Know How to Protect Its Networks." *Wired*, February 1, 2012.

Alcock, John. *The Triumph of Sociobiology*. Oxford: Oxford University Press, 2001.

Arquilla, John. "Twenty Years of Cyberwar." In "Cyberwar and Ethics." Special issue, *Journal of Military Ethics* 12, no. 1 (April 17, 2013): 80–87.

Bailey, Kathleen. "Why the United States Rejected the Protocol to the Biological and Toxin Weapons Convention." Fairfax VA: National Institute for Public Policy, October 2002.

Bell, Stephen. "Cut-Price Stuxnet Successors Possible: Kaspersky." CSO *Magazine*, March 28, 2011.

Bellamy, Alex. *Just Wars*. Cambridge: Polity Press, 2006.

Bernstein, Barton. "The Atomic Bombings Reconsidered." *Foreign Affairs* 74, no. 1 (1995): 135–52.

Bialer, Uri. *The Shadow of the Bomber: The Fear of Air Attack and British Politics, 1932–1939*. London: Royal Historical Society, 1980.

Blasko, Dennis J. "PLA Force Structure: A 20-Year Retrospective." In *Seeking Truth from Facts: A Retrospective on Chinese Military Studies in the Post-Mao Era*, edited by James C. Mulvenon and Andrew N. D. Yang, 51–83. Santa Monica CA: RAND, 2001.

Blechman, Barry, ed. *Unblocking the Road to Zero: Perspectives of Advanced Nuclear Nations*. Washington DC: Stimson Center, April 2009.

Bronk, Christopher, and Eneken Tikk-Ringas. "The Cyber Attack on Saudi Aramco." *Survival: Global Politics and Strategy* 55 (April–May 2013): 81–96.

Brower, Charles F. *World War II in Europe: The Final Year*. London: Palgrave Macmillan, 1998.

Brown, Frederic. *Chemical Warfare: A Study in Restraints*. Piscataway NJ: Transaction Publishers, 2005.

Brown, Gary, and Keira Poellet. "The Customary International Law of Cyberspace." *Strategic Studies Quarterly* (Fall 2012): 126–45.

Buckley, John. *Air Power in the Age of Total War*. London: UCL Press, 1999.

Bull, Hedley. *The Anarchical Society*. New York: Columbia University Press, 1977.

Burns, J. F. "Soviet Peace Charade Is Less than Convincing." *New York Times*, May 16, 1982.

Burton, Kelly. "The Conficker Worm." *SANS*, October 23, 2008.

Campen, Alan D. "Rush to Information-Based Warfare Gambles with National Security." *SIGNAL*, July 1995.

Carpenter, Will D., and Michael Moodie. "Industry and Arms Control." In *Biological Warfare: Modern Offense and Defense*., ed. Raymond A. Zilinskas, 177–91. Boulder CO: Lynne Rienner, 2000.

Carr, Jeffrey. *Inside Cyber Warfare: Mapping the Cyber Underworld*. Sebastopol CA: O'Reilly, 2009.

Carus, Seth W. "Defining 'Weapons of Mass Destruction.'" Center for the Study of Weapons of Mass Destruction, Occasional Paper Series, January 2006.

Checkel, Jeffrey T. "Institutions, and National Identity in Contemporary Europe." *International Studies Quarterly* 13, no. 1 (1999): 83–114.

Clarke, I. F. *Voices Prophesying War.* Oxford: Oxford University Press, 1966.

Clarke, Richard A., and Robert Knake. *Cyber War: The Next Threat to National Security and What to Do about It.* New York: HarperCollins, 2010.

Cliff, Roger, et al. *Entering the Dragon's Lair: The Implications of Chinese Antiaccess Strategies.* Santa Monica CA: RAND, 2007.

Cohen, Eliot A. "The Meaning and Future of Air Power." *Orbis* 39, no. 2 (1995): 189–200.

Cole, Christopher, and E. F. Cheesman. *The Air Defence of Britain, 1914–1918.* London: Bodley Head, 1984.

Cole, Leonard. "The Poison Weapons Taboo: Biology, Culture, and Policy." *Politics and the Life Sciences* 17, no. 2 (1998): 119–32.

Croddy, Eric. *Weapons of Mass Destruction: An Encyclopedia of Worldwide Policy, Technology, and History.* Santa Barbara CA: ABC-CLIO, 2004.

Cyber Conflict Studies Association. "Implication for an Estonia-Like Cyber Conflict for the Government and the Private Sector." Paper presented at Cyber Conflict Studies Association's Annual Symposium, Georgetown University, February 26, 2008.

Dando, Malcolm. *Preventing Biological Warfare: The Failure of American Leadership.* London: Palgrave, 2002.

Davis Biddle, Tami. "Air Power." In *The Laws of War: Constraints on Warfare in the Western World*, edited by Michael Howard, George J. Andreopoulos, and Mark L. Shulman, 140–60. New Haven CT: Yale University Press, 1994.

———. *Rhetoric and Reality in Air Warfare: The Evolution of British and American Ideas about Strategic Bombing, 1914–1945.* Princeton NJ: Princeton University Press, 2002.

DeLamater, John, "The Social Control of Sexuality." *Annual Review of Sociology* 7 (1981): 263–90.

Denning, Dorothy E. *Information Warfare and Security.* Reading MA: Addison-Wesley, 1999.

De Syon, Guillaume. *Zeppelin! Germany and the Airship, 1900–1939.* Baltimore: Johns Hopkins University Press, 2001.

Douglass, Mary. *Purity and Danger: An Analysis of the Concepts of Pollution and Taboo.* New York: Ark, 1984.

Douhet, Giulio. *The Command of the Air.* Washington DC: Office of Air Force History, 1983.

Elster, Jon. "Social Norms and Economic Theory." *Journal of Economic Perspectives* 3, no. 4 (1989): 99–117.

Entous, Adam, and Nathan Hodge. "Oil Firms Hit by Hackers from China, Report Says." *Wall Street Journal*, February 10, 2011.

Finnemore, Martha. "Cultivating International Cyber Norms." In *America's Cyber Future: Security and Prosperity in the Information Age*, chapter 6. Washington DC: Center for a New American Security, June 2011.

Finnemore, Martha, and Kathryn Sikkink. "International Norm Dynamics and Political Change." *International Organization* 52, no. 4 (1998): 887–917.

FireEye Inc. *World War C: Understanding Nation-State Motives behind Today's Advanced Cyber Threats.* Milpitas CA: FireEye, 2013.

Florini, Ann. "The Evolution of International Norms." In "Evolutionary Paradigms in the Social Sciences." Special issue, *International Studies Quarterly* 40, no. 3 (1996): 363–89.

Freedman, Lawrence. "Disarmament and Other Nuclear Norms." *Washington Quarterly* (Spring 2013): 93–108.

Frey, Karsten. "Nuclear Weapons as Symbols: The Role of Norms in Nuclear Policy Making." Paper presented at the Institute Barcelona d'Estudis Internacionals, March 2006.

Gaddis, John. "The Origins of Self-Deterrence: The United States and the Non-Use of Nuclear Weapons, 1945–1958." In *The Long Peace: Inquiries into the History of the Cold War*, 104–46. Oxford: Oxford University Press, 1989.

Gady, Franz-Stefan, and Greg Austin. *Russia, the United States, and Cyber Diplomacy: Opening the Doors.* Washington DC: EastWest Institute, 2010.

Garcia, Denise. *Small Arms and Security: New Emerging International Norms.* New York: Routledge, 2006.

Gärtner, Heinz, Adrian G. V. Hyde-Price, and Erich Reiter. *Europe's New Security Challenges*. Boulder CO: Lynne Rienner, 2001.

Gartzke, Erik. "The Myth of Cyberwar: Bringing War in Cyberspace Back Down to Earth." *International Security* 38, no. 2 (2013): 41–73.

Gavin, James. *War and Peace in the Space Age*. New York: Harper Brothers, 1958.

Genest, Marc. *Conflict and Cooperation: Evolving Theories of International Relations. Recent Security Developments*. Boston: Wadsworth, 2004.

Glenn, John. "Realism versus Strategic Culture: Competition and Collaboration?" *International Studies Review* 11, no. 3 (2009): 523–51.

Godwin, Paul H. B. "Compensating for Deficiencies: Doctrinal Evolution in the Chinese People's Liberation Army: 1978–1999." In *Seeking Truth from Facts: A Retrospective on Chinese Military Studies in the Post-Mao Era*, edited by James C. Mulvenon and Andrew N. D. Young, 87–118. Santa Monica CA: RAND, 2001.

Goertz, Gary. *International Norms and Decision Making: A Punctuated Equilibrium Model*. Lanham MD: Rowman and Littlefield, 2003.

Goldschmidt, B. "The Negotiation of the Non-Proliferation Treaty (NPT)." *IAEA Bulletin*, 22, nos. 3–4 (1980): 73–80.

Gorman, Siobhan, and Julian Barnes. "Cyber Combat: Act of War." *Wall Street Journal*, May 31, 2011.

Grotius, Hugo. *The Law of War and Peace*. Translated by Francis Kelsey. Indianapolis IN: Bobbs-Merrill, 1925.

Guenter, Scot. "Stockholm Peace Appeal." In *W. E. B. Du Bois: An Encyclopedia*, edited by Gerald Horne and Mary Young, 201–2. Westport CT: Greenwood Press, 2001.

Harris, Robert, and Jeremy Paxman. *A Higher Form of Killing: The Secret Story of Chemical and Biological Warfare*. New York: Hill and Wang, 1982.

Hathaway, Melissa. "Cyber Readiness Index 1.0." Paper. Science, Technology, and Public Policy Program, Belfer Center for Science and International Affairs, Kennedy School, Harvard University, November 8, 2013.

Healey, Jason. *A Fierce Domain: Conflict in Cyberspace, 1986 to 2012*. Washington DC: Atlantic Council and Cyber Conflict Studies Association, 2013.

Hecht, Jeff. "Top 10: Weapons of the Future." *New Scientist*, September 4, 2006.

Heickerö, Roland. "Emerging Cyber Threats and Russian Views on Information Warfare and Information Operations." Stockholm: Swedish Defence Research Agency, 2010.

Hiro, Dilip. *The Longest War: The Iran-Iraq Military Conflict*. New York: Routledge, 1991.

Hofmann, Claudia, and Ulrich Schneckener. "Special Report: NGOs and Nonstate Armed Actors, Improving Compliance with International Norms." Washington DC: United States Institute of Peace, July 2011.

Holloway, David. *Stalin and the Bomb: The Soviet Union and Atomic Energy 1939–1956*. New Haven CT: Yale University Press, 1994.

Horowitz, Michael. *The Diffusion of Military Power: Causes and Consequences for International Politics*. Princeton NJ: Princeton University Press, 2012.

Houqing, Wang, and Xingye Zhang, eds. *Science of Campaigns*. Beijing: Beijing National Defense University Press, 2000.

Hudson, Manley O. "Present Status of the Hague Conventions of 1899 and 1907." *American Journal of International Law* 25, no. 1 (1931): 114–17.

Hugo, Victor. *The History of a Crime*. Whitefish MT: Kessinger, 2004.

Hurley, Alfred F. *Billy Mitchell: Crusader for Air Power*. Bloomington: Indiana University Press, 2006.

International Council of Chemical Associations. "Statement on the 10th Anniversary of the Chemical Weapons Convention." April 27, 2007.

International Group of Experts. *Tallinn Manual on the International Law Applicable to Cyber Warfare*. Edited by Michael N. Schmitt. Cambridge: Cambridge University Press, 2013.

Jones, H. A. *The War in the Air*. Oxford: Clarendon Press, 1937.

Kahn, Herman. "Nuclear Proliferation and Rules of Retaliation." *Yale Law Journal* 76, no. 1 (1966): 77–91.

———. *Thinking about the Unthinkable*. New York: Touchstone, 1985.

Karsh, Efraim. *The Iran-Iraq War: 1980–1988*. Oxford: Osprey, 2002.

Kartchner, Kerry. "Weapons of Mass Destruction and the Crucible of Strategic Culture." Comparative Strategic Cultures Curriculum. Ft. Belvoir VA: Defense Threat Reduction Agency, October 31, 2006.

Katzenstein, Peter, Alexander Wendt, and Ronals Jepperson. "Norms, Identity, and Culture in National Security." In *The Culture of National Security: Norms and Identity in World Politics*, edited by Peter Katzenstein, 33–75. New York: Columbia University Press, 1996.

Keegan, John. *The Second World War*. New York: Viking, 1989.

Kello, Lucas. "The Meaning of the Cyber Revolution: Perils to Theory and Statecraft." *International Security* 38, no. 2 (2013): 7–40.

Kennedy, John. "Press Conference, March 21, 1963." In *Public Papers of the Presidents of the United States: John F. Kennedy*, 280–81. Washington DC: U.S. Government Printing Office, 1964.

Kennett, Lee. *A History of Strategic Bombing: From the First Hot-Air Balloons to Hiroshima and Nagasaki*. New York: Charles Scribner's Sons, 1982.

Keohane, Robert, and Joseph Nye. *Power and Interdependence*. New York: Pearson Scott Foresman, 1989.

———. *Power and Interdependence: World Politics in Transition*. Boston: Little, Brown, 1977.

Klimburg, Alexander. *National Cyber Security Framework Manual*. Tallinn, Estonia: NATO CCD COE, 2012.

Knake, Robert. "Testimony on Planning for the Future of Cyber Attack Attribution." U.S. House of Representatives, Committee on Science and Technology, July 15, 2010.

Koblentz, Gregory. *Living Weapons: Biological Warfare and International Security*. Ithaca NY: Cornell University Press, 2009.

Koblentz, Gregory, and Brian Mazanec. "Viral Warfare: The Security Implications of Cyber and Biological Weapons." *Comparative Strategy* 32, no. 5 (2013): 418–34.

Kopp, Carlo. "Moore's Law and Its Implications for Information Warfare." Paper presented at the Third International AOC EW Conference, January 6, 2002.

Leahy, William D. *I Was There*. New York: McGraw-Hill, 1950.

Lee, Robert M. "The Interim Years of Cyberspace." *Air and Space Power Journal* (January–February 2013): 58–79.

Legro, Jeffrey W. "Military Culture and Inadvertent Escalation in World War II." *International Security* 18, no. 4 (1994): 108–42.

———. "Which Norms Matter? Revisiting the 'Failure' of Internationalism." *International Organization* 51, no. 1 (1997): 31–63.

Lewis, James. "Confidence-Building and International Agreement in Cybersecurity." *Disarmament Forum* 4 (2011): 51–59.

Lewis, James, and Katrina Timlin. *Cybersecurity and Cyberwarfare: Preliminary Assessment of National Doctrine and Organization*. Washington DC: Center for Strategic and International Studies, 2011.

Liang, Qiao, and Wang Xiangsui. *Unrestricted Warfare*. Beijing: PLA Literature and Arts Publishing House, 1999.

Liff, Adam P. "Cyberwar: A New 'Absolute Weapon'? The Proliferation of Cyberwarfare Capabilities and Interstate War." *Journal of Strategic Studies* 35, no. 3 (2012): 401–28.

Longmate, Norman. *The Bombers: The RAF Offensive against Germany 1939–1945*. London: Arrow, 1988.

Lorell, Mark A. *The U.S. Combat Aircraft Industry, 1909–2000: Structure, Competition, Innovation*. Santa Monica CA: RAND, 2003.

Lyons, Marty. "Threat Assessment of Cyber Warfare: A White Paper." University of Washington, December 7, 2005.

Mandelbaum, Michael. *The Nuclear Question: The United States and Nuclear Weapons: 1946–1976*. Cambridge: Cambridge University Press, 1979.

———. *The Nuclear Revolution: International Politics before and after Hiroshima*. Cambridge: Cambridge University Press, 1981.

Matthew, Richard, and Kenneth Rutherford. "The Evolutionary Dynamics of the Movement to Ban Landmines." *Alternatives: Global, Local, Political* 28, no. 1 (2003): 29–56.

Maurer, Tim. "Cyber Norm Emergence at the United Nations: An Analysis of the Activities at the UN Regarding Cyber-Security." Belfer Center for Science and International Affairs, Kennedy School, Harvard University, September 2011.

Medalia, Jonathan. "Nuclear Testing and Comprehensive Test Ban: Chronology Starting September 1992." *Congressional Research Service*, 97–1007 F (November 9, 2004).

Moravcsik, Andrew. "The Origins of Human Rights Regimes." *International Organization* 54 (2000): 217–52.

Morgenthau, Hans. "Politics among Nations." In *International Relations and Political Theory*. Vancouver: UBC Press, 1993.

Murray, Williamson. "A Tale of Two Doctrines: The Luftwaffe's Conduct of the Air War, and the USAF's Manual I-I." *Journal of Strategic Studies* 6, no. 4 (1983): 84–93.

Norris, John. *Early Gunpowder Artillery: 1300–1600*. Marlborough UK: Crowood Press, 2003.

Norris, Robert. *Racing for the Bomb: General Leslie Groves, the Manhattan Project's Indispensible Man*. South Royalton VT: Steerforth Press, 2002.

Nye, Joseph, and John D. Donahue, eds. *Governance in a Globalizing World*. Washington DC: Brookings Institution Press, 2000.

Ochsner, Herman. *The History of German Chemical Warfare in World War II, Part 1: The Military Aspect*. Fort Belvoir VA: Historical Office of the Chief of the Chemical Corps, 1949.

O'Dwyer, Diana. "First Landmines, Now Small Arms? The International Campaign to Ban Landmines as a Model for Small Arms Advocacy." *Irish Studies in International Affairs* 17 (2006): 77–97.

Offices of U.S. Congressmen Markey and Waxman. "Electric Grid Vulnerability: Industry Responses Reveal Security Gaps." U.S. House of Representatives, May 21, 2013.

Organization for the Prohibition of Chemical Weapons Technical Secretariat. "Note by the Technical Secretariat: Support by Inspected States Parties for Sampling and Analysis under Article VI of the Chemical Weapons Convention." S/548.2006, February 10, 2006.

———. "Status of Participation in the Chemical Weapons Convention as of 14 October 2013." S/1131/2013, October 14, 2013.

Osgood, Robert. *NATO: The Entangling Alliance*. Chicago: University of Chicago Press, 1962.

Overy, Richard J. *The Air War: 1939–1945*. Washington DC: Potomac Books, 2005.

Pape, Robert. *Bombing to Win: Air Power and Coercion in War.* Ithaca NY: Cornell University Press, 1996.

Parks, W. Hays. "Air War and the Law of War." *Air Force Law Review* 32, no. 1 (1990): 1–225.

Parlow, Anita. "Banning Land Mines." *Human Rights Quarterly* 16, no. 4 (1994): 715–39.

Paul, T. V. *The Tradition of Non-Use of Nuclear Weapons.* Stanford CA: Stanford Security Studies, January 2009.

Pickering, William H. "The Future of Artificial Flight." *Aeronautics* 6, no. 2 (1908): 16–17.

Price, Richard. *The Chemical Weapons Taboo.* Ithaca NY: Cornell University Press, 1997.

———. "A Genealogy of the Chemical Weapons Taboo." *International Organization* 49, no. 1 (1995): 73–103.

———. "Reversing the Gun Sights: Transnational Civil Society Targets Land Mines." *International Organization* 52, no. 3 (1998): 613–44.

Puchala, Donald, and Stuart Fagan. "International Politics in the 1970s: The Search for a Perspective." *International Organization* 28 (1974): 247–66.

Ratner, Steven. "International Law: The Trials of Global Norms." In "Frontiers of Knowledge." Special issue, *Foreign Policy* 110 (Spring 1998): 65–80.

Rauscher, Karl Frederick, and Valery Yaschenko. *Russia-U.S. Bilateral on Cybersecurity: Critical Terminology Foundations.* Washington DC: EastWest Institute, 2011.

Rearden, Steven L. *History of the Office of the Secretary of Defense,* vol. 1: *The Formative Years, 1947–1950.* Washington DC: U.S. Government Printing Office, 1984.

Renz, Bettina, and Sibylle Scheipers. "Discrimination in Aerial Bombing: An Enduring Norm in the 20th Century?" *Defence Studies* 12, no. 1 (2012): 17–43.

Rid, Thomas."Cyber War Will Not Take Place." *Journal of Strategic Studies* 35, no. 1 (2011): 5–32.

———. *Cyber War Will Not Take Place.* Oxford: Hurst/Oxford University Press, 2012.

Rose, Gideon. "Neoclassical Realism and Theories of Foreign Policy." *World Politics* 51, no. 1 (1998): 144–72.

Rotblat, Joseph. "Leaving the Bomb Project." *Bulletin of the Atomic Scientists* 41, no. 7 (1985): 16–19.

Rublee, Maria Rost. *Nonproliferation Norms: Why States Choose Nuclear Restraint.* Athens: University of Georgia Press, 2009.

Rumble, Greville. *The Politics of Nuclear Defence—A Comprehensive Introduction.* Cambridge: Polity Press, 1985.

Ryngaert, Cedric. "Working Paper: Non-State Actors and International Humanitarian Law." Catholic University of Leuven, Faculty of Law, 2008.

Sagan, Scott. *Ethics and Weapons of Mass Destruction: Religious and Secular Perspectives.* Cambridge: Cambridge University Press, 2004.

———. "Why Do States Build Nuclear Weapons? Three Models in Search of a Bomb." *International Security* 21, no. 3 (1996–1997): 54–86.

Sanger, David E., and Steven Erlanger. "Suspicion Falls on Russia as 'Snake' Cyberattacks Target Ukraine's Government." *New York Times*, March 8, 2014.

Schindler, Dietrich, and Jiri Toman. *The Laws of Armed Conflict.* 3rd ed. Dordrecht: Martinus Nijhoff, 1988.

Schmitt, Michael N. "International Law in Cyberspace: The Koh Speech and Tallinn Manual Juxtaposed." *Harvard International Law Journal* 54 (December 2012): 17–32.

Scott, James Brown. *The Proceedings of the Hague Peace Conference (1899).* Oxford: Oxford University Press, 1920.

Shoham, Dany. "Chemical and Biological Weapons in Egypt." *Nonproliferation Review* (Spring–Summer 1998): 48–58.

Shultz, George, William Perry, Henry Kissinger, and Sam Nunn. "Toward a Nuclear-Free World." *Wall Street Journal*, January 15, 2008.

———. "A World Free of Nuclear Weapons." *Wall Street Journal*, January 4, 2007.

Singer, P. W., and Allan Friedman. *Cybersecurity and Cyberwar: What Everyone Needs to Know.* Oxford: Oxford University Press, 2014.

Spaight, J. M. *Air Power and War Rights.* Harlow UK: Longmans, Green, 1924.

Spitzer, Kirk. "A Forgotten Horror: The Great Tokyo Air Raid." *Time*, March 27, 2012.

Swyter, Hans. "Political Considerations and Analysis of Military Requirements for Chemical and Biological Weapons." *Proceedings of the National Academies of Sciences* 65, no. 1 (1970): 261–70.

Tannenwald, Nina. *The Nuclear Taboo: The United States and the Non-Use of Nuclear Weapons since 1945.* Cambridge: Cambridge University Press, 2008.

———. "Stigmatizing the Bomb: Origins of the Nuclear Taboo." *International Security* 29, no. 4 (2005): 5–49.

Tannenwald, Nina, and Richard Price. "Norms and Deterrence: The Nuclear and Chemical Weapons Taboos." In *The Culture of National Security: Norms and Identity in World Politics*, ed. Peter J. Katzenstein, 114–52. New York: Columbia University Press, 1996.

Tarzi, Shah. "The Role of Principles, Norms and Regimes in World Affairs." *International Journal on World Peace* 15, no. 4 (1998): 5–27.

Taylor, Telford. *Final Report to the Secretary of the Army on the Nuremberg War Crimes Trials under Control Council Law No. 10.* Washington DC: U.S. Government Printing Office, 1949.

Thomas, Ward. *The Ethics of Destruction: Norms and Force in International Relations.* Ithaca NY: Cornell University Press, 2001.

Tkacik, John. "Trojan Dragon: China's Cyber Threat." *Heritage Foundation Backgrounder*, February 8, 2008.

Tucker, Jonathan. "A Farewell to Germs: The U.S. Renunciation of Biological and Toxin Warfare, 1969–70." *International Security* 27, no. 1 (2002): 107–48.

———. *War of Nerves: Chemical Warfare from World War I to Al-Qaeda.* New York: Random House, 2007.

United Nations General Assembly. A/68/98, "Report of the Group of Governmental Experts on Developments in the Field of Information and

Telecommunications in the Context of International Security," June 24, 2013.
United Nations Office for Disarmament Affairs. "Fact Sheet: Developments in the Field of Information and Telecommunications in the Context of International Security," June 2013.
United Nations Security Council. "Report of the Mission Dispatched by the Secretary-General to Investigate Allegations of the Use of Chemical Weapons in the Conflict between the Islamic Republic of Iran and Iraq." S/17911, March 21, 1986.
United States. "National Intelligence Estimate 100-05-55: Implications of Growing Nuclear Capabilities for the Communist Bloc and the Free World," June 14, 1955.
U.S. Department of Defense. "The Definition of Cyberspace." Deputy Secretary of Defense Memorandum, May 12, 2008.
———. *Joint Publication 1-02: Department of Defense Dictionary of Military and Associated Terms*. Washington DC: Joint Chiefs of Staff, May 15, 2011.
U.S. Department of Energy. *The Manhattan Project: Making the Atomic Bomb*. Washington DC: Department of Energy, 1999.
U.S. Department of State. *Conference on the Limitation of Armament*. Washington DC: U.S. Government Printing Office, 1922.
U.S. General Accounting Office. "B-2 Bomber: Status of Cost, Development, and Production." GAO/NSIAD-95-164, August 1995.
U.S. Government Accountability Office. "Defense Department Cyber Efforts: Definitions, Focal Point, and Methodology Needed for DOD to Develop Full-Spectrum Cyberspace Budget Estimates." GAO-11-695R, July 29, 2011.
U.S. National Research Council. *Technology, Policy, Law, and Ethics Regarding U.S. Acquisition and Use of Cyberattack Capabilities*. Washington DC: National Academies Press, 2009.
U.S. Secretary of State. "International Security Advisory Board Task Force: Draft Report on China's Strategic Modernization." September 2008.
U.S. War Department. *The War of the Rebellion: A Compilation of the Official Records of the Union and Confederate Armies*. Ser. III, vol. 3. Washington DC: U.S. Government Printing Office, 1899.

Verdirame, Guglielmo. "Testing the Effectiveness of International Norms: UN Humanitarian Assistance and Sexual Apartheid in Afghanistan." *Human Rights Quarterly* 23, no. 3 (2001): 733–68.

Verne, Jules. *Clipper of the Clouds*. London: Searle and Rivington, 1887.

———. *Master of the World*. In *Works of Jules Verne*, vol. 14. London: F. Tyler Daniels, 1911.

Wansbrough, Henry, ed. *The New Jerusalem Bible*. New York: Doubleday, 1985.

Webber, Liz. "Computer Use Expected to Top 2 Billion." *Inc. Magazine*, July 2, 2007.

Webster, Charles Kingsley, and Noble Frankland. *The Strategic Air Offensive against Germany: 1939–1945*. Vol. 4. *London: Her Majesty's Stationery Office*, 1961.

Wegener, Henning. "Cyber Peace." In *The Quest for Cyber Peace*. Geneva: International Telecommunication Union and World Federation of Scientists, 2011.

Wells, H. G. *The World Set Free*. London: W. Collins Sons, 1924.

Wendt, Alexander. *Social Theory of International Politics*. Cambridge: Cambridge University Press, 1999.

Wise, S. F. *Canadian Airmen and the First World War*. Toronto: University of Toronto Press, 1980.

Wittner, Lawrence S. *Resisting the Bomb: A History of the World Nuclear Disarmament Movement, 1954–1970*. Stanford CA: Stanford University Press, 1997.

Woolley, Pamela. "Defining Cyberspace as a United States Air Force Mission." Wright-Patterson Air Force Base OH: Air Force Institute of Technology, June 2006.

Wright, Robin. *Dreams and Shadows: The Future of the Middle East*. New York: Penguin Press, 2008.

INDEX